PHYSICS RESEARCH AND TECHNOLOGY

# UNDERSTANDING HEAT CONDUCTION

# PHYSICS RESEARCH AND TECHNOLOGY

Additional books and e-books in this series can be found
on Nova's website under the Series tab.

PHYSICS RESEARCH AND TECHNOLOGY

# UNDERSTANDING HEAT CONDUCTION

## WILLIAM KELLEY
### EDITOR

Copyright © 2021 by Nova Science Publishers, Inc.

**All rights reserved.** No part of this book may be reproduced, stored in a retrieval system or transmitted in any form or by any means: electronic, electrostatic, magnetic, tape, mechanical photocopying, recording or otherwise without the written permission of the Publisher.

We have partnered with Copyright Clearance Center to make it easy for you to obtain permissions to reuse content from this publication. Simply navigate to this publication's page on Nova's website and locate the "Get Permission" button below the title description. This button is linked directly to the title's permission page on copyright.com. Alternatively, you can visit copyright.com and search by title, ISBN, or ISSN.

For further questions about using the service on copyright.com, please contact:
Copyright Clearance Center
Phone: +1-(978) 750-8400          Fax: +1-(978) 750-4470          E-mail: info@copyright.com.

### NOTICE TO THE READER

The Publisher has taken reasonable care in the preparation of this book, but makes no expressed or implied warranty of any kind and assumes no responsibility for any errors or omissions. No liability is assumed for incidental or consequential damages in connection with or arising out of information contained in this book. The Publisher shall not be liable for any special, consequential, or exemplary damages resulting, in whole or in part, from the readers' use of, or reliance upon, this material. Any parts of this book based on government reports are so indicated and copyright is claimed for those parts to the extent applicable to compilations of such works.

Independent verification should be sought for any data, advice or recommendations contained in this book. In addition, no responsibility is assumed by the Publisher for any injury and/or damage to persons or property arising from any methods, products, instructions, ideas or otherwise contained in this publication.

This publication is designed to provide accurate and authoritative information with regard to the subject matter covered herein. It is sold with the clear understanding that the Publisher is not engaged in rendering legal or any other professional services. If legal or any other expert assistance is required, the services of a competent person should be sought. FROM A DECLARATION OF PARTICIPANTS JOINTLY ADOPTED BY A COMMITTEE OF THE AMERICAN BAR ASSOCIATION AND A COMMITTEE OF PUBLISHERS.

Additional color graphics may be available in the e-book version of this book.

## Library of Congress Cataloging-in-Publication Data

ISBN: 978-1-53619-182-0

*Published by Nova Science Publishers, Inc. † New York*

# CONTENTS

| | | |
|---|---|---|
| **Preface** | | vii |
| **Chapter 1** | Cooling Kinetics in Stone Fruits<br>*Francisco Javier Cuesta and*<br>*María Dolores Alvarez* | 1 |
| **Chapter 2** | Sensitivity of Numerical Modeling Technique for Conjugate Heat Transfer Involving High Speed Compressible Flow over a Cylinder<br>*Laurie A. Florio* | 73 |
| **Chapter 3** | Advances in Heat Conduction Analysis with Fundamental Solution Based Finite Element Methods<br>*Qing-Hua Qin* | 123 |
| **Index** | | 165 |

# PREFACE

The first chapter of this book proposes an analytical Fourier series solution to the equation for heat transfer by conduction in a spherical shell with an internal stone consisting of insulating material as a model for the kinetic of temperature in stone fruits both as a general solution and a mass average value. The chapter also considers an internal heat source linearly reliant on temperature. The second chapter focuses on the sensitivity of the numerical modeling technique for conjugate heat transfer involving high speed compressible flow over a cylinder. The last chapter presents an overview of the fundamental solution (FS) based finite element method (FEM) and its application in heat conduction problems. First, basic formulations of FS-FEM are presented, such as the nonconforming intra-element field, auxiliary conforming frame field, modified variational principle, and stiffness equation. Then, the FS-FE formulation for heat conduction problems in cellular solids with circular holes, functionally graded materials, and natural-hemp-fiber-filled cement composites are described.

Chapter 1 - Stone fruits, like plums, cherries, olives and so forth, have approximately spherical or ellipsoid geometries, but inside they contain a ligneous core -the seed- whose physical and thermo-physical parameters are radically different from those of the edible part, the pulp. Moreover, the contact surface between this seed and the pulp is in practice the deepest point that can be reached in the fruit and it performs the role of a "thermal centre" which in homogeneous solid objects is performed by the geometric centre. There are a number of important issues that need to be known in food chilling, for example the time required to reach a given temperature at the thermal centre of the product, or how much heat has to be extracted to reach that temperature, or what is the average temperature of the product once it has reached a given temperature at its thermal centre or on the surface, etc. This chapter proposes an analytical Fourier series solution to the equation for heat transfer by conduction in a spherical shell with an internal stone consisting of insulating material as a model for the kinetic of temperature in stone fruits both a general solution and a mass average value. The chapter also considers an internal heat source linearly reliant on temperature. The first approximations to the general equations are used to derive the model equations for calculation of the above mentioned issues and also for indirect calculations of Biot number, or the thermal diffusivity when the Biot number is given.

Chapter 2 - The surface heating due to conjugate heat transfer is often the main source of conduction heat transfer through solid materials moving through or impinged upon by high speed compressible flows. High temperature gradients and large heat fluxes at the fluid-solid interfaces develop under such conditions. Since high temperatures and high temperature gradients affect the solid material properties and thus material strength, an understanding of these conjugate heat transfer phenomena is important. The compressible flow around a fixed cylinder or due to a moving cylinder is a well-established baseline for comparison. Numerical models can be used to investigate and predict the resulting thermal conditions for such flow and heat conduction effects, but the numerical modeling methods must be attuned to properly capture the conditions unique to this type of flow. Among the factors influencing the numerical results are the mesh size, the time step and time discretization, various discretization methods including gradient calculations and gradient limiters as well as the turbulence model and associated parameters and options. When a moving object requiring some mesh motion is added to the simulation, additional options become available. A sliding type of mesh motion treats a particular zone as a rigid body, moving the cylinder along a linear path, for quadrilateral/hexahedral element types. Remeshing algorithms deform elements and then split or coalesce elements as an object moves through the mesh, generally requiring a triangular/tetrahedral element type that may be more numerically diffuse. An overset mesh, where the mesh associated with the cylinder moves over a fixed background mesh, can be used to model the movement of the cylinder in a simulation with quadrilateral/hexahedral elements, without remeshing. However, the accuracy of the overset method for high temperature gradient conditions is not well defined. The sensitivity of conditions that develop as a result of high speed compressible flow over a cylinder to changes in the mesh, numerical models, and mesh motion methods is explored in this work with comparisons made to published data. All studies are conducted for a 0.0762m diameter steel cylinder in air with the pressure, temperature, and heat flux around the outer surface of the cylinder compared for the various models. A Mach 6.46 flow is studied first with comparisons made to published data. Then, the sensitivity of the flow conditions to the modeling methods or model related parameters is investigated for the same system. In the final part of the study, three different methods of applying mesh motion are implemented for subsonic, near sonic, and supersonic flow conditions to examine the effect of the mesh motion method on the predicted flow conditions and the relationship to the flow speed. These studies provide the information needed to better select the modeling methods to use for conjugate heat transfer analysis with high speed compressible flow under a given set of conditions.

Chapter 3 - This chapter presents an overview of the fundamental solution (FS) based finite element method (FEM) and its application in heat conduction problems. First, basic formulations of FS-FEM are presented, such as the nonconforming intra-element field, auxiliary conforming frame field, modified variational principle, and stiffness equation. Then, the FS-FE formulation for heat conduction problems in cellular solids with circular holes, functionally graded materials, and natural-hemp-fiber-filled cement composites are described. With this method, a linear combination of the fundamental solution at different points is actually used to approximate the field variables within the element. Meanwhile, the independent frame field defined along the elemental boundary and the modified variational functional are employed to guarantee inter-element continuity as well as to generate the final stiffness equation and establish the linkage between the boundary frame field and the internal field in the element. Finally, a brief summary of the approach is provided and future trends in this field are identified.

In: Understanding Heat Conduction
Editor: William Kelley

ISBN: 978-1-53619-182-0
© 2021 Nova Science Publishers, Inc.

*Chapter 1*

# COOLING KINETICS IN STONE FRUITS

### *Francisco Javier Cuesta and María Dolores Alvarez**
Institute of Food Science, Technology and Nutrition (ICTAN-CSIC),
Ciudad Universitaria, Madrid, Spain

### ABSTRACT

Stone fruits, like plums, cherries, olives and so forth, have approximately spherical or ellipsoid geometries, but inside they contain a ligneous core -the seed- whose physical and thermo-physical parameters are radically different from those of the edible part, the pulp. Moreover, the contact surface between this seed and the pulp is in practice the deepest point that can be reached in the fruit and it performs the role of a "thermal centre" which in homogeneous solid objects is performed by the geometric centre. There are a number of important issues that need to be known in food chilling, for example the time required to reach a given temperature at the thermal centre of the product, or how much heat has to be extracted to reach that temperature, or what is the average temperature of the product once it has reached a given temperature at its thermal centre or on the surface, etc. This chapter proposes an analytical Fourier series solution to the equation for heat transfer by conduction in a spherical shell with an internal stone consisting of insulating material as a

---

* Corresponding Author's E-mail: mayoyes@ictan.csic.es.

model for the kinetic of temperature in stone fruits both a general solution and a mass average value. The chapter also considers an internal heat source linearly reliant on temperature. The first approximations to the general equations are used to derive the model equations for calculation of the above mentioned issues and also for indirect calculations of Biot number, or the thermal diffusivity when the Biot number is given.

**Keywords**: stone fruit, transient heat transfer, Fourier series, heat of respiration

# General Introduction:
# Some Concepts in Heat Transfer

As it is well known, heat conduction through a food is governed by the Fourier equation. In addition, in solids with simple shaped geometries (infinite flat plate, infinite circular cylinder or sphere), with no heat generation in them and with no phase change throughout the limits of the thermal process, the Fourier equation may be written as described in references [1-5]:

$$\frac{1}{r^\Gamma}\frac{\partial}{\partial r}\left(k \cdot r^\Gamma \cdot \frac{\partial T}{\partial r}\right) = \rho c \frac{\partial T}{\partial t} \qquad (1)$$

Where $T$ is the temperature, $t$ the time, $r$ the distance to the origin, $\rho$ the density, $c$ the specific heat per unit mass and $k$ the thermal conductivity. $\Gamma$ is a geometrical constant whose value is $\Gamma = 0$ for the flat plate, $\Gamma = 1$ for the infinite circular cylinder y $\Gamma = 2$ for the sphere.

This fundamental law describes the temperature dynamic as a function depending on external geometry, external temperature and surrounding medium, etc. Furthermore, the heat flow depends on the thermo-physical parameters of food (thermal conductivity, specific heat per unit mass and density), which in turn are functions of the chemical composition of the food and the temperature. In itself, the Fourier equation represents the law of conservation of energy, the mathematical description of the energy balance at a specific point into the food. The solution to this equation depends on the shape and size of the food, its internal homogeneity or heterogeneity, whether there is a phase change inside it and the way the heat is removed or added to the food. That is, it depends on its thermophysical properties and the boundary conditions.

If the solid is homogeneous and isotropic with constant thermal parameters, the conductivity can be extracted from the derivative, and the Fourier equation remains:

$$\frac{k}{r^\Gamma}\frac{\partial}{\partial r}\left(r^\Gamma \cdot \frac{\partial T}{\partial r}\right) = \rho c \frac{\partial T}{\partial t} \qquad (2)$$

Which, dividing by $k$ in both sides of the equation, remains:

$$\frac{1}{r^\Gamma}\frac{\partial}{\partial r}\left(r^\Gamma \cdot \frac{\partial T}{\partial r}\right) = \frac{\rho c}{k}\frac{\partial T}{\partial t} \qquad (3)$$

The most commonly used boundary condition is the Newton equation, which relates the heat flow through the surface to the difference between the surface temperature and the medium temperature:

$$-k\left[\frac{\partial T}{\partial r}\right]_{r=R} = h(T_{sf} - T_{ex}) \qquad (4)$$

The "minus" sign means that the heat is extracted from the solid. $R$ is the minor semi length of the solid, $h$ the surface heat transfer coefficient, $T_{sf}$ the temperature at the surface of the body and $T_{ex}$ the medium temperature.

In addition to the previous boundary condition, we must add the condition imposed by the internal symmetry of the food:

$$\left[\frac{\partial T}{\partial r}\right]_{r=0} = 0 \qquad (5)$$

And it will be considered as initial condition that the temperature is uniform in the object. That is:

$$t = 0; T(r) = T_0 = Cte \tag{6}$$

Eq. (3) may be written:

$$\frac{\partial^2 T}{\partial r^2} + \frac{\Gamma}{r}\frac{\partial T}{\partial r} = \frac{1}{\alpha}\cdot\frac{\partial T}{\partial t} \tag{7}$$

Where the thermal diffusivity of the food is:

$$\alpha = \frac{k}{\rho c}$$

Therefore, Eq. (4) remains:

$$\left[\frac{\partial T}{\partial r}\right]_{r=R} = -\frac{h}{k}\left(T_{sf} - T_{ex}\right) \tag{8}$$

In Eqs. (5-8), distance $r$ is measured in m, temperature $T$ in °C, time $t$ in s, thermal conductivity $k$ in J/(s m K), surface heat transfer coefficient $h$ in J/(s m² K), specific heat $c$ en J/(kg K) and the density $\rho$ in kg/m³. Thermal diffusivity $a$ is then measured in m²/s.

With the change to dimensionless variables:

$$Y = \frac{T - T_{ex}}{T_0 - T_{ex}} \tag{9}$$

$$x = \frac{r}{R} \tag{10}$$

$$Fo = \frac{\alpha \cdot t}{R^2} \tag{11}$$

$$Bi = \frac{h \cdot R}{k} \tag{12}$$

Eq. (7) with external boundary condition (8), internal symmetry condition (5) and initial condition (6) remains:

$$\frac{\partial^2 Y}{\partial x^2} + \frac{\Gamma}{x}\frac{\partial Y}{\partial x} = \frac{\partial Y}{\partial Fo} \tag{13}$$

$$\left[\frac{\partial Y}{\partial x}\right]_{x=1} = -BiY_{sf} \tag{14}$$

$$\left[\frac{\partial Y}{\partial x}\right]_{x=0} = 0 \tag{15}$$

$$Fo = 0; Y = 1 \tag{16}$$

$Y$ is the relative difference of temperatures, $x$ the relative internal coordinate, $Fo$(Fourier number) the dimensionless time and $Bi$ (Biot number) ratio between internal thermal resistance of the food and the thermal resistance of the cooling medium.

It is also well known [1-5] that the solution to Eq. (13) with conditions (14) to (16) is given by the following sum of infinite terms (Fourier series):

$$Y = \sum_{n=1}^{\infty} A_n \psi(\delta_n x) e^{-\delta_n^2 Fo} \tag{17}$$

In accordance with Cuesta et al. [10]:

$$A_n = \frac{2Bi}{\psi(\delta_n)[\delta_n^2 + Bi^2 - (\Gamma-1)Bi]} \tag{18}$$

As mentioned above, the parameter $\Gamma$ takes values 0, 1 or 2, respectively, in the cases of the infinite slab, infinite cylinder or sphere. $\psi(\delta_n x) = \cos(\delta_n x)$ for an infinite slab, $\psi(\delta_n x) = J_0(\delta_n x)$ for an infinite cylinder and $\psi(\delta_n x) = \frac{\sin(\delta_n x)}{\delta_n x}$ for a sphere. $\delta_n$ are the solutions to the Biot equation (14) for boundary conditions:

$$\left[\frac{\partial \psi(\delta_n x)}{\partial x}\right]_{x=1} = -Bi \cdot \psi(\delta_n) \tag{19}$$

## Estimations and Applications

### *Cooling/Heating Times*

The roots $\delta_n$ derived from Eq. (19) are discrete values increasing with the terms of the series, so that from a certain value onwards only the first term in the series is significant and Eq. (17) may be replaced with sufficient accuracy by its first addend:

$$Y \approx A_1 \psi(\delta_1 x) e^{-\delta_1^2 Fo}$$

Or, regardless of its index:

$$Y \approx A \psi(\delta x) e^{-\delta^2 Fo} \tag{20}$$

At the thermal centre $[\psi(\delta x)]_{x=0} = 1$ and Eq. (20) remains:

$$Y \approx A e^{-\delta^2 Fo} \tag{21}$$

Solving for $Fo$ in Eq. (20):

$$Fo \approx \frac{\ln(A\psi(\delta x)/Y)}{\delta^2} \tag{22}$$

And at the centre:

$$Fo \approx \frac{\ln(A/Y)}{\delta^2} \tag{23}$$

For the average temperature in the food, Eq. (17) must be averaged out for volume, leaving as follows:

$$\bar{Y} = \sum_{n=1}^{\infty} \bar{A}_n e^{-\delta_n^2 Fo} \tag{24}$$

where

$$\bar{A}_n = A_n \bar{\psi}(\delta_n) = \frac{2(\Gamma-1)Bi^2}{\delta_n^2[\delta_n^2+Bi^2-(\Gamma-1)Bi]} \quad (25)$$

And, as in the previous case, when only the firs addend is retained:

$$\bar{Y} \approx \bar{A}e^{-\delta^2 Fo} \quad (26)$$

Eqs. (21) and (22) are very important due to their different applications. For example, in food refrigeration processing design, since 1965 these equations allow estimating cooling or heating times at the centre of the food [6-14]. In addition, they are necessary in the indirect measure of important parameters as surface heat transfer coefficient of the refrigerant medium or in the indirect measure of the thermo-physical parameters of the food [15-17]. In turn, these parameters are necessary to be accurately predicted, as they are basic control parameters to design the postharvest processing, necessary to reduce losses and to extend the shelf life of the food [18].

### *Example I (from Reference [5])*

The issue is to cool apples using a flow of air at 5°C, from an initial temperature of 20°C down to a temperature of 8°C in the centre [5]. The surface heat transfer coefficient is h = 10 W/m²K and the thermo-physical properties of the apple are: conductivity k = 0.4 W/m.K, specific heat C = 3.8 kJ/kg.K and density ρ = 960 kg/m³. Calculation of the processing time is requested.

### **Solution**

As the author himself indicates [5], the apples may be likened to spheres with a diameter 2R = 8 cm.

Consequently, the thermal diffusivity is: $\alpha = 0.4/(960 \times 3.8 \times 10^3) = 1.0965 \times 10^{-7} \, m^2/s$

The radius is: $R = 0.08/2 = 0.04$ m

As the shape is considered as spherical: $\Gamma = 2$

The Biot number (Eq. (12)): $Bi = \frac{10 \times 0.04}{0.4} = 1$

The first $\delta$ value corresponding to $Bi = 1$ is: $\delta = \pi/2$

And the first $A$ value (Eq. (18)): $A = \frac{2 \times 1}{\frac{\sin(\pi/2)}{\pi/2}[(\pi/2)^2 + 1^2 - (2-1) \times 1]} = 1.273$

In this case (Eq. (9)): $Y = \frac{8-5}{20-5} = 0.2$

Applying Eq. (22): $Fo \approx \frac{\ln\left(\frac{A}{Y}\right)}{\delta^2} = \frac{\ln\left(\frac{1.273}{0.2}\right)}{\left(\frac{\pi}{2}\right)^2} = 0.75$

Finally, solving for $t$ in Eq. (11): $t = \frac{Fo \times R^2}{\alpha} = \frac{0.75 \times 0.04^2}{\alpha} = 10947$ s $= 3.04$ h

This first section has been focused on Eqs. (2) to (6), which are the simplest on heat transfer. However, there are many options in heat transfer that are not described by these equations and that are nevertheless necessary to describe other options (for instance irregular shape, non internal homogeneity of the thermal parameters, anisotropy, etc.).

In present chapter, three types of modifications to these equations are introduced: firstly (in the next second section) the internal inhomogeneity of the fruit considering the presence of the stone inside; secondly (in the third section), the internal source of heat produced by its respiration; and thirdly (in the fourth section), the joint modification produced by both at the same time.

# Modelling Thermal Kinetics in Stone Fruits

In most cases, to calculate the parameters seen in the above section, the first approximation to the general Fourier series solution are applied considering solid geometries of homogeneous isotropic bodies, which usually does not explicitly include the stone. In addition, the indirect measurements of thermal diffusivity and the external heat transfer coefficient are also based on these approximations [15, 16, 19]. Therefore, it would be useful to have appropriate mathematical models that explicitly include the stone, to allow accurate prediction of the design parameters mentioned above, and to measure the thermal diffusivity and the film coefficient.

As mentioned above, next section will be devoted to model in analytical Fourier series the solution to the equation for heat transfer by conduction in a stone fruit, which will be assimilated to a spherical shell with its internal stone consisting of insulating material. In a later section will be included the internal heat of respiration.

## Mathematical Background

Stone fruits, such as olives, cherries, plums, and so forth, have approximately spherical or ellipsoid geometries, but inside they contain a ligneous core – the seed –, whose physical and thermo-physical parameters are radically different from those of the edible part, the pulp. Moreover, the contact surface between this seed and the pulp is, in practice, the deepest point that can be reached in the fruit and it performs the role of a "thermal centre" which in homogeneous solid objects is represented by the geometric centre.

Consequently, consider the stone fruit as a sphere of radius $R$ containing a concentric spherical kernel of radius $a$ (Figure 1), which is to be considered quasi-insulating against the effects of heat transfer, as compared to external heat transfer, for the duration of the process. It was decided to adopt this condition in this chapter as a consequence of having found practically no values for the thermo-physical parameters of the drupe's stone, and the fact that, in the cases examined by the authors, the stone has a woody layer with a porous appearance, in many cases externally similar to cork, protecting the seed inside. Therefore, it will be assumed that the initial temperature of the object, $T_0$, is uniform, and that it is placed in sudden contact with a mechanically stirred medium at a uniform temperature $T_{ex}$.

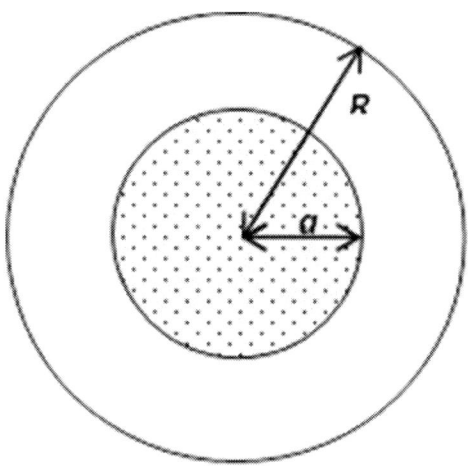

Figure 1. Schematic of the model.

The Fourier equation (Eq. (2)) is still valid, as does the boundary equation (Eq. (4)). Instead the internal boundary equation (Eq. (5)) becomes the following:

$$\left[\frac{\partial T}{\partial r}\right]_{r=r_0} = 0 \tag{27}$$

Thus, with the above change to dimensionless variables (Eqs. (9-12)), the system to be solved is the one formed by Eqs. (13), (14) and (27) with the initial condition (16):

$$\frac{\partial^2 Y}{\partial x^2} + \frac{\Gamma}{x}\frac{\partial Y}{\partial x} = \frac{\partial Y}{\partial Fo}$$

$$\left[\frac{\partial Y}{\partial x}\right]_{x=1} = -BiY_{sf}$$

$$\left[\frac{\partial Y}{\partial x}\right]_{x=x_0} = 0$$

with:

$$x_0 = \frac{r_0}{R} \tag{28}$$

and

$$Fo = 0; Y(x) = 1$$

The general solution can be expressed as in the previous case with Eq. (17) [13]:

$$Y(x, Fo) = Y = \sum_{n=1}^{\infty} A_n \psi(\delta_n x) e^{-\delta_n^2 Fo} \tag{29}$$

And its mass average value:

$$\bar{Y} = \sum_{n=1}^{\infty} \overline{A_n} e^{-\delta_n^2 Fo} \tag{30}$$

But in this case $\psi(\delta_n x)$ is:

$$\psi(\delta_n x) = \frac{\sin[\delta_n(x-x_0)] + \delta_n x_0 \cos[\delta_n(x-x_0)]}{\delta_n x} \tag{31}$$

$\delta_n$ are the eigenvalues of the transcendental boundary equation and $A_n$ are the constants of the series expansion, whose values in the general case are, respectively [13]:

$$\delta_n = \tan[\delta_n(1-x_0)] \frac{1-Bi+\delta_n^2 x_0}{1+x_0(Bi-1)} \tag{32}$$

$$A_n = \frac{2Bi\psi(\delta_n)}{[\psi(\delta_n)]^2(\delta_n^2+Bi^2-Bi)-\delta_n^2 x_0^3} \tag{33}$$

When $Bi \to \infty$ Eqs. (32) and (33) become:

$$\tan[\delta_{n,M}(1-x_0)] + \delta_{n,M}x_0 = 0 \tag{34}$$

$$A_{n,\infty} = \frac{-2}{\cos[\delta_{n,M}(1-x_0)]} \frac{1}{(1+\delta_{n,M}^2 x_0^2 - \delta_{n,M}^2 x_0^3)} \tag{35}$$

and

$$\psi(\delta_{n,M}) = \frac{\sin[\delta_{n,M}(1-x_0)] + \delta_{n,M} \cdot x_0 \cos[\delta_{n,M}(1-x_0)]}{\delta_{n,M}} = 0 \tag{36}$$

Where the case $Bi \to \infty$ is identified by the subscript $\infty$, except for $\delta_{n,M}$, which is identified by the subscript "M" to indicate that it is the maximum value that can be achieved (for $Bi \to \infty$). As can be seen, these equations explicitly include the relative coordinate of the nucleus. Furthermore, Eqs. (31) to (36) reduce to those of the solid sphere when $x_0 = 0$.

For the mass average the constants are [13]:

$$\bar{A}_n = A_n \bar{\psi}(\delta_n) = \frac{6Bi^2}{\delta_n^2(1-x_0^3)} \cdot \frac{[\psi(\delta_n)]^2}{[\psi(\delta_n)]^2[\delta_n^2+Bi^2-Bi]-\delta_n^2 x_0^3} \tag{37}$$

$$\bar{A}_{n,\infty} = A_{n,\infty}\bar{\psi}(\delta_{n,M}) = \frac{6}{\delta_{n,M}^2(1-x_0^3)} \frac{1+\delta_{n,M}^2 x_0^2}{(1+\delta_{n,M}^2 x_0^2 - \delta_{n,M}^2 x_0^3)} \tag{38}$$

Figure 2 shows the first eigenvalue ($\delta_1$) of boundary Eq. (32) vs. $Bi$ for $x_0 = 0.3$ to $0.6$, and solid sphere as reference. The reason for this range is that, according to Cinquanta et al. [20], Hernández et al. [21], and Jiménez-Jiménez et al. [22], it is the average diameter ratio range of the stone. These values were confirmed in an extensive work on prevention of browning of two varieties of table olives by pre-cooling in cold water, carried out at the Institute of Food Science, Technology and Nutrition

(ICTAN-CSIC). Practical examples of the data are provided later in this chapter.

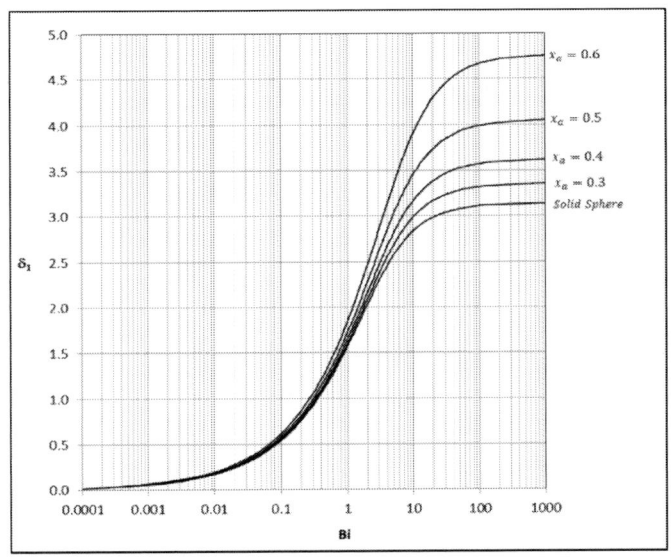

Figure 2. First eigenvalue ($\delta_1$) of boundary equation (32) vs. $Bi$ for $x_a = 0.3$ to 0.6, and solid sphere as reference.

The heat flow transferred by the fruit through the entire surface at moment $t$ is (Newton's law):

$$\dot{Q} = -4\pi R^2 h\left(T_{sf} - T_{ex}\right) = -4\pi R^2 h \Delta T_0 \sum A_n \psi(\delta_n) e^{-\delta_n^2 Fo}$$

Or, denoting by $\dot{Q}_0$ the total initial heat flow transferred by the fruit through its surface:

$$\dot{Q}_0 = 4\pi R^2 h \Delta T_0 \tag{39}$$

$$\dot{Q} = -\dot{Q}_0 \sum A_n \psi(\delta_n) e^{-\delta_n^2 Fo} \tag{40}$$

As a result, the total energy extracted from the fruit up to moment $t$ (i.e., $Fo$) is (by integrating Eq. (40) and rearranging terms):

$$Q(t) = \int_0^t \dot{Q} dt' = \dot{Q}_0 \sum \frac{A_n}{\delta_n^2} \psi(\delta_n)\left(e^{-\delta_n^2 Fo} - 1\right) \qquad (41)$$

The average flow transferred by the fruit until time $t$ is:

$$\bar{Q}(t) = \frac{1}{t}\int_0^t \dot{Q} dt' = \frac{\dot{Q}_0}{Fo} \sum \frac{A_n}{\delta_n^2} \psi(\delta_n)\left(e^{-\delta_n^2 Fo} - 1\right) \qquad (42)$$

The maximum amount of energy removable from the fruit ($t = \infty$) at $Bi = Cte$, and with the initial temperature difference $\Delta T_0$, is:

$$Q_M = \frac{\dot{Q}_0}{a} R^2 \sum \frac{A_n}{\delta_n^2} \psi(\delta_n) \qquad (43)$$

And the absolute maximum amount of heat extractable from the fruit for $t = \infty$ and $Bi = \infty$ is:

$$Q_{M,Bi \to \infty} = \frac{K}{a} R^2 \sum \frac{A_{n,\infty}}{\delta_{n,M}^2} \frac{-1}{\sin[\delta_{n,M}(1-x_0)]}$$

## Estimations and Applications

### *Cooling/Heating Times*

As in the case of an homogeneous solid sphere, from a given moment onwards the infinite series converges very quickly, because from the second exponential factor onwards values also decreases very quickly, so in practice the contribution to the final sum of the complete series is null. Thus, for practical purposes, exact analytical solutions lead to first-approximation models based on a linear approximation of the cooling graph (on a semi-logarithmic scale), which are valid from a given time. Apart from estimating the cooling/heating times, the main applications of these linear approximations are the indirect measurement of thermo-physical parameters (thermal diffusivity and conductivity) and the surface heat transfer coefficient. In dimensionless terms, the first term of the complete series at the thermal centre (stone–pulp contact) is:

$$Y(Fo) = Y \approx A_1 e^{-\delta_1^2 Fo} \qquad (44)$$

where $A_1$ is the lag factor and $\delta_1$ the first eigenvalue of the boundary condition (Eq. (32)).

As in the solid sphere, the time required to reach a dimensionless temperature $Y_0$ at the stone–pulp interface is:

$$Fo_{Y_0} \approx \frac{\ln\left[\frac{A_1}{Y_0}\right]}{\delta_1^2} \qquad (45)$$

In the case of mass average temperature, $A_1$ should be substituted by $\bar{A}_n = A_n \bar{\psi}(\delta_n)$ (Eq. (37)).

If $Bi \to \infty$, the slope attains its maximum $(-\delta_{1,M}^2)$. Therefore. the dimensionless time required to reach $Y_0$ is minimized:

$$Fo_{Y_a,min} \approx \frac{\ln(A_{1,\infty}/Y_0)}{\delta_{1,M}^2} \qquad (46)$$

As in the case of $\delta_{1,M}$, when $Bi \to \infty$ the dimensionless cooling/heating time ($Fo_{Y_0,min}$) is indicated by the subscript "*min*" to emphasize that it is the absolute minimum dimensionless time necessary to achieve the dimensionless temperature difference $Y_0$, as the surface heat transfer coefficient is considered to be infinite.

### *Thermal Flow*

In the same manner as in Eq. (44) for the kinetic temperature, the estimation of the heat flow and the amount of energy removed from the fruit up to moment $t$ can be approximated by the first term of Eqs. (40) and (41):

$$\dot{Q} \approx -\dot{Q}_0 A_1 \psi(\delta_1) e^{-\delta_1^2 Fo} \qquad (47)$$

And

$$Q(t) \approx \dot{Q}_0 \frac{A_1}{\delta_1^2} \psi(\delta_1) \left(e^{-\delta_1^2 Fo} - 1\right)$$

$$\dot{Q}_0 = 4\pi R^2 h \Delta T_0$$

And the time average power needed to cool the fruit:

$$\bar{\dot{Q}}(t) \approx \frac{\dot{Q}_0}{Fo} \frac{A_1}{\delta_1^2} \psi(\delta_1) \left(e^{-\delta_1^2 Fo} - 1\right)$$

As at the stone–pulp interface (Eq. (44)):

$$Y_0 \approx A_1 e^{-\delta_1^2 Fo}$$

And the average power needed to cool the fruit up to the dimensionless temperature difference $Y$ can be approximated as:

$$\bar{\dot{Q}}(t) \approx \dot{Q}_0 \psi(\delta_1) \frac{A_1 - Y}{\ln A_1 - \ln Y}$$

## *Indirect Measurement of Thermal Diffusivity and Surface Heat Transfer Coefficient*

According to Cuesta et al. [10], Erdoğdu [16, 19], Erdoğdu et al. [18], and Uyar and Erdoğdu [24], the experimental lag factor $A_1$ allows us to determine the thermal diffusivity and the Biot number. In fact, the time-temperature history of the product at the thermal centre (stone–pulp contact) can be transformed into a Table $\{Y_i, t_i(s)\}$, where $Y_i = (T_i - T_{ex})/(T_0 - T_{ex})$, and therefore the linear regression of the linear portion of its graph in semi-logarithmic coordinates can be obtained. In fact, Eq. (44) can also be written as:

$$\ln Y \approx \ln A_1 - \delta_1{}^2 Fo = \ln A_1 - S \cdot t \qquad (48)$$

$$S \equiv \frac{a \cdot \delta_1{}^2}{R^2} \qquad (49)$$

This $A_1$ obtained from the linear regression can be named as $A_{1,exp}$ and, by identifying with Eq. (49) and resolving, $a$ is deduced as well:

$$a = \frac{SR^2}{\delta_1^2} \qquad (50)$$

The experimental $\delta_1$ can be determined by a trial and error method, and therefore the Biot number (that is, the dimensionless surface heat transfer coefficient) and the thermal diffusivity can also be determined. In fact, from a trial value $\delta_1$, the Biot number that corresponds to this trial $\delta_1$ can be deduced by resolving $Bi$ in Eq. (32):

$$Bi = 1 + \frac{\delta_1^2 x_0 - m}{1 + m x_0} \qquad (51)$$

where:

$$m = \frac{\delta_1}{\tan[\delta_1(1-x_0)]} \qquad (52)$$

And taking into account Eq. (31) and Eq. (33) for $n = 1$ and $x = 1$:

$$\psi(\delta_1) = \frac{\sin[\delta_1(1-x_0)] + \delta_1 x_0 \cos[\delta_1(1-x_0)]}{\delta_1} \qquad (53)$$

$$A_1 = \frac{2Bi\psi(\delta_1)}{[\psi(\delta_1)]^2[\delta_1^2 + Bi^2 - Bi] - \delta_1^2 x_0^3} \qquad (54)$$

If $A_1 = A_{1,exp}$ (within the deviation considered), then $\delta_1$, $Bi$ (i.e., the surface heat transfer coefficient) and the thermal diffusivity $a$ (by substituting in Eq. (50)) have been determined simultaneously, and therefore can be regarded as experimental.

If $A_1 \neq A_{1,exp}$ (within the deviation considered) then, by increasing or decreasing $\delta_1$, $A_1$ can be recalculated until the desired accuracy is obtained.

In practice, using a simple algorithm to accelerate the convergence, no more than 4 or 5 cycles of calculation are needed to determine $\delta_1$, $Bi$, and $a$.

## *Example II (from Reference [13])*

The purpose of this demonstrative example is to illustrate the application of equations deduced in the preceding section, in the two different kinds of applications:

First: determination of thermal diffusivity and surface heat transfer coefficient. This requires measuring the temperature kinetics (Table $\{t(s) - T(°C)\}$).

Second: Estimation of design parameters (cooling/heating time, processing equipment). In this case, we have to know the thermo-physical parameters of the fruit and the surface heat transfer coefficient.

### Experiment Description

In an experiment on prevention of browning of table olives by pre-cooling in cold water, the fruits were cooled in order to evaluate the residual effect of the water pre-cooling on the rate of browning in a storage chamber at 4°C. Previously, the fruits were stabilized at 17°C, and then dipped in a 4-litre bath with a mixture of water and crushed ice, stirred by a agitator. At the time of introducing the olive samples, this mixture was stabilized at 0.4°C, approximately.

One of the fruits above had a K-type bare wire thermocouple probe in contact with the stone–pulp interface and there were another K-type bare wire thermocouple probe measuring the temperature of the water.

The olive sample weighed 10.44 g, with an equatorial diameter $D = 23.25$ mm and a longitudinal diameter $L = 32.44$ mm. After the complete experiment, the stone was measured. Its dimensions were as follows:

weight 1.95 g, equatorial diameter $d = 10.02$ mm, and longitudinal diameter $l = 21.95$ mm.

**Equivalent Sphere**

In order to reduce the shape of the sample to the model considered in this chapter, the olive was considered as a sphere having the same volume as the actual olive. As it is very approximately an elongated ellipsoid, the radius of the equivalent sphere is:

$$R = \frac{1}{2}\sqrt[3]{D^2 L} = 0.013 \text{ m}$$

And, similarly, the radius of the sphere equivalent to the stone is:

$$r_0 = \frac{1}{2}\sqrt[3]{d^2 l} = 0.0065 \text{ m}$$

Hence, the ratio of radii is:

$$x_0 = 0.501$$

And the density of the pulp of the olive:

$$\rho = 1057.6 \text{ kg/m}^3$$

**Determination of Biot Number and Thermal Diffusivity**

Figure 3a shows the semi-logarithmic-scale plot of the experimental values {Y, t (s)}, its linear portion, and the regression line of the linear zone. The lag factor $A_1$ and the cooling kinetic corresponding to $Bi = \infty$ as the absolute limit case, are also shown.

The linear regression is:

$$\ln Y = 0.2921 - 6.323 \; 10^{-3} t$$

with

$$R^2 = 0.999$$

Therefore the experimental lag factor is:

$$A_{1,exp} = e^{0.2921} = 1.339$$

and the absolute slope:

$$S = \frac{\delta_1^2 a}{R^2} = 6.323\ 10^{-3} s^{-1}$$

By applying the iterative process described above (Eqs. (51) to (54)), parameters $\delta_1$ and $Bi$ are simultaneously obtained, and, by substituting $S$ in Eq. (50), the thermal diffusivity is determined:

$$\delta_1 = 2.93$$

$$Bi = 4.46$$

$$a = 1.239\ 10^{-7}\ m^2 s^{-1}$$

From the chemical composition of the olive, the thermal diffusivity can be theoretically estimated according to ASHRAE [25], Carson [26], Carson et al. [27, 28], Choi and Okos [29], Marcotte et al. [30], Maroulis et al. [31], and Murakami and Okos [32], and compared with the diffusivity that has just been determined.

The chemical composition of the olive used for this chapter is as follows: $x_{Water} = 75.97\%$; $x_{Prot} = 4.37\%$; $x_{fat} = 10.23\%$; $x_{Ash.} = 0.91\%$; $x_{Carboh} = 8.52\%$ (estimated by difference) (averaged from Guillén et al. [33], ASHRAE [25], Ongen et al. [34] and Wang et al. [35]).

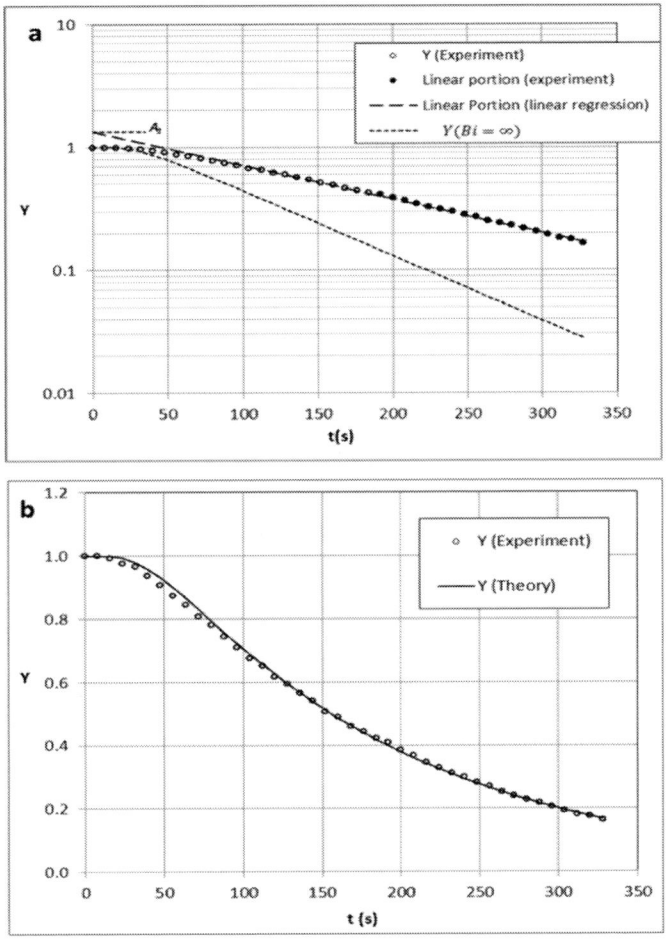

Figure 3. (a): Semi-logarithmic-scale plot of the experimental values {Y,t (s)} (hollow circles), linear portion (black circles) and the regression line of the linear zone (unbroken line), lag factor $A_1$ and cooling kinetic corresponding to $Bi = \infty$ as the absolute limit case (dotted line). (b): Complete theoretical series of the cooling process (unbroken line) compared with the experimental values (empty circles).

Thus, the following values are calculated, which can be compared with those just obtained: thermal diffusivity $a = 1.268\ 10^{-7}\ m^2/s$ (2.4%); density $\rho = 1038.2\ kg/m^3$ (1.8%). By accepting the theoretical specific heat per unit mass $c_p = 3608.6\ J/kg\ K$, the conductivity can be derived from the determined diffusivity:

$$k = \rho c_p a = 1057.6 \times 3608.6 \times 1.239 \, 10^{-7} = 0.4730 \text{ W/m K}$$

This can be compared with that obtained from the chemical composition: $k = 0.4752$ W/m K (deviation 0.5%). In consequence, the estimated surface heat transfer coefficient is:

$$h = \frac{k}{R} Bi = 162.3 \text{ W/m}^2 \text{ K}$$

On the other hand, as at this point the Biot number is already known ($Bi = 4.46$), the roots of the transcendental boundary equation (Eq. (32)) and the constants of the serial expansion (Eq. (33)) can be calculated. Therefore, the complete theoretical time-temperature curve of this experiment can be drawn and compared with the experimental values. The result is shown in Figure 3b (unbroken line), together with the experimental values (empty circles). The standard error between the experimental and theoretical $Y$ curves is $\pm 0.01$ (1.16%).

### *Asymptotic Aproximation to Dimensionless Slope $\delta_1^2$*

In practice, the first eigenvalue ($\delta_1$) of the transcendental Eq. (32) is one of the most important parameter in the process, as knowledge of it allows us to identify the surface heat transfer coefficient. Therefore, $-\delta_1^2$ is the absolute slope of the linear portion of the cooling/heating curve in semi-logarithmic scale:

$$\delta_1^2 = -\frac{d(\ln Y)}{d(Fo)}$$

*Asymptote $Bi \to 0$:*

As seen in Figure 2, the plot of $\delta_1$ vs. $Bi$ is similar to that of a solid sphere (also drawn) but higher in value. When $Bi \to 0$, $\delta_1 \to 0$ as well, and their values can be estimated by considering the expressions derived from Cuesta and Alvarez [13]:

$$\delta_1{}^2 \approx \frac{M_1{}^2}{(1-x_0)} \cdot \frac{Bi}{Bi(1-x_0)(M_1+2x_0-2)+x_0 M_1{}^2+2(1-x_0)^2} \qquad (55)$$

with:

$$M_1 = (\pi/2)^2$$

The maximum deviation of this equation is around 4% for $Bi \leq 5$ and $0 \leq x_0 \leq 0.9$. The standard deviation is 1.44%. For $Bi \leq 1$ the maximum deviation is $\leq 2.5\%$, the standard deviation being $< 0.7\%$.

Eq. (55) is also valid for $x_0 = 0$ (solid sphere):

$$\delta_1{}^2 \approx (\pi/2)^4 \frac{Bi}{Bi[(\pi/2)^2-2]+2} \qquad (56)$$

Approximation (56) greatly improves the approximation found by Cuesta and Lamúa [11], which for the sphere can be written:

$$\delta_1{}^2 \approx 6 \frac{Bi}{Bi+2} \qquad (57)$$

Indeed, this approximation (56) leads to a deviation $< 3\%$ for $Bi \leq 5$ and to a deviation $< 1\%$ for $Bi < 3$. In the case of Eq. (57), they only had these deviations for $Bi < 0.25$.

**Maximum Values of $\delta_n$**

The maximum slope $\delta_{1,M}^2$ depends only on the shape of the object. In our case, because the geometry is spherical, it depends only on the size of the stone. Furthermore, the values $\delta_{1,M}^2$ can be approximated as in the above case for each particular value of $x_0$ by considering the expressions derived from a previous study [13]. The approximate expression is as follows:

$$\delta_{n,M}^2 \approx \frac{1}{(1-x_0)^2} \times \frac{M_n^2}{M_n - 2(1-x_0)} \tag{58}$$

Or:

$$\delta_{n,M} \approx \frac{1}{(1-x_0)} \times \frac{M_n}{\sqrt{M_n - 2(1-x_0)}} \tag{59}$$

with:

$$M_n = \left[\frac{(2n-1)\pi}{2}\right]^2 \tag{60}$$

For $n = 1$, $M_1 = (\pi/2)^2$ and:

$$\delta_{1,M}^2 \approx \frac{1}{(1-x_0)^2} \times \frac{(\pi/2)^4}{(\pi/2)^2 - 2(1-x_0)} \tag{61}$$

Or:

$$\delta_{1,M} \approx \frac{1}{(1-x_0)} \times \frac{(\pi/2)^2}{\sqrt{(\pi/2)^2 - 2(1-x_0)}} \tag{62}$$

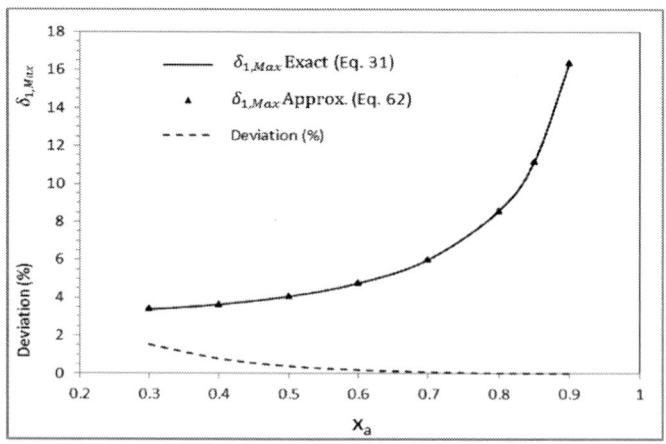

Figure 4. Comparison of $\delta_{1,Max}$ obtained with the approximation (62) and the exact value (equation (31)). It is also shown the deviation (%).

Eq. (55) includes Eq. (61) for $Bi \to \infty$.

Figure 4 shows $\delta_{1,M}$ obtained with approximation (62), the exact value (Eq. (34)), and the deviation (%).

### Example III. Prediction of Cooling Times in Example II

An important issue in postharvest handling is the time that the fruit must stay in the hydro-cooler to achieve a certain temperature at the stone–pulp interface before moving to the next phase of its conservation or treatment. Thus, in this example the cooling time is estimated by Eq. (45), but approximating the slope by Eq. (55). It is assumed that the surface heat transfer coefficient ($h$), the initial temperature of the fruit ($T_0$), the water temperature in the hydro-cooler ($T_{ex}$), and the nutritional composition of the fruit are known.

For the reason that this is merely an illustrative example, we take the same case in reverse as in the *Example II*, and we could enunciate it as follows: in a water tank in which $T_{ex} = 0.46$ °C, the olive previously described in the Experiment Description of the above application (*Example II*) has to be cooled from the initial temperature $T_0 = 17.7$ °C to the final temperature $T = 4$ °C at the stone–pulp interface ($Y = 0.205$). The surface heat transfer coefficient is $h \approx 162$ W/m² K. How long does it take to reach that temperature? What is the average temperature at that moment?

As seen in the Equivalent sphere subsection, the radius of the equivalent sphere is $R = 0.013$ m, the ratio of the stone radius to the pulp radius is $x_a = 0.501$ and, as seen in Example II, from the nutritional composition the theoretical thermo- physical parameters are: thermal diffusivity $a = 1.268 \, 10^{-7}$ m²/s, density $\rho = 1038.2$ kg/m³, specific heat per unit mass $c_p = 3608.6$ J/kg K, and conductivity $k = 0.4752$ W/m K. As the surface heat transfer coefficient is $h = 162$ W/m² K, the Biot number is:

$$Bi = \frac{162 \cdot 0.013}{0.4752} = 4.43$$

Consequently, by approximating the slope with Eq. (55) and by introducing this value in Eqs. (31) and (33):

$$\delta_1^2 \approx 7.95 \rightarrow \psi(\delta_1) = 0.432 \text{ and } A_1 = 1.155$$

And by substitution in Eq. (45):

$$Fo_{Y_a} \approx 0.217$$

Therefore, the estimated time (in seconds) is:

$$t_{Y_a} \approx \frac{R^2}{a} Fo_Y = 296 \text{ s}$$

The time deduced from the linear regression is $t = 297\ s$. Consequently, the deviation is –0.23%.

The mass average temperature of the pulp at that moment is (by replacing $A_1$ in Eq. (44) by $\bar{A}_1 = A_1 \bar{\psi}$ (Eq. (37)), rearranging terms, and clearing $\bar{Y}_a$:

$$\bar{Y}_a \approx \bar{A}_1 \exp(-\delta_1^2 Fo_{Y_a}) \approx 0.1$$

That is:

$$\bar{T} = T_{ex} + \bar{Y}_a (T_0 - T_{ex}) \approx 2\ °C$$

The estimated power required to cool the olive would be (Eqs. (39) and (40)):

$$\dot{Q}_0 = 4\pi \cdot (0.013)^2 \cdot 162 \cdot 17.24 = 5.922\ W$$

$$\bar{\dot{Q}}(t) \approx 1.40\ W\ (6\%\ \text{deviation from complete series})$$

Therefore, considering that this specimen of Gordal olive weighed 10.44 g, the average calorific value would be 134.6 W/Kg, and consequently for a load of 1 Tm (for example) the value would be: $\bar{Q}(t) \approx$ 135 kW.

# MODELLING THERMAL KINETICS CONSIDERING INTERNAL LINEARLY TEMPERATURE DEPENDENT HEAT GENERATION

As previously said, the second alteration of the Fourier equations, which is introduced in this chapter, consists on considering the internal source of heat produced by the respiration of the fruit.

In addition to the homogeneity and isotropy conditions, until now it has been considered that the object to be cooled/heated is a "passive" object, that is, no heat is generated inside.

The heat transfer by conduction through temperature-dependent internal heat sources as in fruits and vegetables, where there are additional problems of geometry, packaging, storage, respiration heat, etc., is generally a highly complex problem requiring sophisticated numerical procedures only available on a computer. In this context, respiration heat is generally considered to be a function that increases exponentially with temperature [36-38], but various authors use other models in practical cases. For instance, some authors take heat generation to be a constant value [39-42]; then, others treat it as a constant in theory but in practice consider it to be negligible in comparison with other more powerful heat sources [43]. Other authors take a potential model [37, 38, 44], or even a function of time [36]. There are regression models like those of Kole and Prasad [45] (regression to a fourth-grade polynomial) or Rao et al. [46] (regression to a sigmoidal model). Exact analytical solutions have been derived for some particular cases, including heat generation. Jakob [1], for example, considered the case of conduction with a linear heat source under steady state conditions and Carslaw & Jaeger [2] proposed a solution to the problem in transient conditions for an infinite slab. A solution was also proposed for an infinite cylinder and infinite Biot number.

Next section is devoted to model in analytical Fourier series the solution to the equation for heat transfer by conduction in a fruit with a linearly temperature-dependent internal heat of respiration source. It will be considered the fruit as solid and having a single dimension, with constant, homogeneous and isotropic thermal parameters.

The reason for adopting a linear rate model for heat generation is that an exact solution is still feasible using separation of variables; moreover, there is an abundance of data in the literature supporting the adoption of such a linear model as a first approximation, for example in Xu and Burfoot [47], who adopt a linear approach in the case of potato chilling.

## Mathematical Background

The one-dimensional Fourier equation for heat transfer in a single dimension, with constant, homogeneous and isotropic coefficients and a linear heat source may be written thus:

$$k \frac{1}{r^\Gamma} \frac{\partial}{\partial r}\left(r^\Gamma \cdot \frac{\partial T}{\partial r}\right) + q_0 + q_1 T = \rho c \frac{\partial T}{\partial t} \tag{63}$$

With the boundary condition of the third kind:

$$-k \left[\frac{\partial T}{\partial r}\right]_{r=R} = h(T_{sf} - T_{ex})$$

As in the previous cases, $\Gamma$ is the geometrical constant ($\Gamma = 0$ for the flat plate, $\Gamma = 1$ for the infinite circular cylinder y $\Gamma = 2$ for the sphere).

Table 1 shows some values of $q_0$ and $q_1$ obtained by linear regression of the average values taken from the ASHRAE [12].

## Table 1. Summary of regression values calculated from ASHRAE® [12] except from*

| Variety | $q_0$ (W kg$^{-1}$) | $q_1$ (W kg$^{-1}$K$^{-1}$) |
|---|---|---|
| Apples "Y transparent" | 0.0097 | 0.0073 |
| Apples Average | 0.0061 | 0.0037 |
| Apples Early Cultivars | 0.0101 | 0.0040 |
| Apples Late Cultivars | 0.0052 | 0.0025 |
| Apricots | 0.0060 | 0.0038 |
| Artichokes Globe | 0.0723 | 0.0124 |
| Asparagus | 0.1040 | 0.0478 |
| Beans Lima Unshelled | 0.0232 | 0.0211 |
| Beans Lima Shelled | 0.0303 | 0.0334 |
| Beans Snap | -0.0153 | 0.0189 |
| Beets Red Roots | 0.0160 | 0.0026 |
| Black Berries | 0.0251 | 0.0213 |
| Blue Berries | 0.0028 | 0.0097 |
| Broccoli | 0.0291 | 0.0506 |
| Brussels Sprouts | 0.0436 | 0.0179 |
| Cabbage Penn State | 0.0080 | 0.0048 |
| Cabbage Red Early | 0.0168 | 0.0070 |
| Cabbage Savoy | 0.0165 | 0.0182 |
| Cabbage White Spring | 0.0235 | 0.0085 |
| Cabbage White Winter | 0.0104 | 0.0045 |
| Cauliflower | 0.0284 | 0.0089 |
| Cauliflower | 0.0298 | 0.0117 |
| Celery | 0.0120 | 0.0060 |
| Gooseberries | 0.0214 | 0.0039 |
| Peas Green | 0.0486 | 0.0400 |
| Plums, Wickson | 0.00503 | 0.00266 |
| Potatoes* | 0.0174 | 0.0019 |
| Potatoes California white, rose | | |
| immature | 0.0140 | 0.0038 |
| mature | 0.0146 | 0.0009 |
| very mature | 0.01225 | 0.0009 |
| Raspberries | -0.0028 | 0.0222 |
| Strawberries | -0.0033 | 0.0213 |
| *From Xu and Burfoot [47] | | |

The above equations may be written in dimensionless form:

$$\frac{1}{x^\Gamma} \frac{\partial}{\partial x}\left(x^\Gamma \cdot \frac{\partial Y}{\partial x}\right) + \alpha^2 Y + \beta = \frac{\partial Y}{\partial Fo} \tag{64}$$

$$\left[\frac{\partial Y}{\partial x}\right]_{x=1} = -Bi \cdot Y_{sf} \tag{65}$$

In Eqs. (64) and (65) the nomenclature is the same as in Eqs. (7-12). Dimensionless numbers $\alpha^2$ and $\beta$ are the two new dimensionless numbers which describes the heat source:

$$\alpha^2 = \frac{q_1 R^2}{k} \tag{66}$$

$$\beta = \frac{(q_0 + q_1 T_{ex})R^2}{k(T_0 - T_{ex})} = \frac{B_0 R^2}{k \Delta T_0} \tag{67}$$

where

$$B_0 = q_0 + q_1 T_{ex} \tag{68}$$

$$\Delta T_0 = T_0 - T_{ex} \tag{69}$$

## General Solution for Simple Geometries

The general solution for Eq. (64) with the boundary condition (65) may be written as follows [48]:

$$Y(x, Fo) - Y_s(x) = \sum_{n=1}^{\infty} A_n \psi(\delta_n x) e^{-(\delta_n^2 - \alpha^2)Fo} \tag{70}$$

where $Y_s$ is the steady solution (see [48]):

$$Y_s(x) = \frac{T_s - T_{ex}}{\Delta T_0} = \frac{\beta}{\alpha^2}\left[\frac{Bi \cdot \psi(\alpha \cdot x)}{\alpha \cdot \psi'(\alpha) + Bi \cdot \psi(\alpha)} - 1\right] \tag{71}$$

The function $\psi$ has the following expressions:

$$\psi(\delta_n x) = \cos(\delta_n x); \ \psi(\alpha x) = \cos(\alpha x) \text{ for an infinite slab}$$

$\psi(\delta_n x) = J_0(\delta_n x); \psi(\alpha x) = J_0(\alpha x)$ for an infinite cylinder

And:

$\psi(\delta_n x) = \dfrac{\sin(\delta_n x)}{\delta_n x}; \psi(\alpha x) = \dfrac{\sin(\alpha x)}{\alpha x}$ for a sphere

Hereafter, as long as there is no risk of confusion, we use the notation:

$\psi_\delta = \psi(\delta)$
$\psi'_\delta = \psi'(\delta) = \left[\dfrac{\partial \psi(\delta x)}{\partial x}\right]_{x=1}$

And:

$\psi_\alpha = \psi(\alpha)$
$\psi'_\alpha = \psi'(\alpha) = \left[\dfrac{\partial \psi(\alpha x)}{\partial x}\right]_{x=1}$

$\delta_n$ are the eigenvalues of the transcendental equation (as in the no heat generation case):

$$\delta \cdot \psi'_\delta = -Bi \cdot \psi_\delta \tag{72}$$

And coefficients $A_n$ (see [48]):

$$A_n = A_{n,0}\left(1 - \dfrac{\beta}{\delta_n^2 - \alpha^2}\right) \tag{73}$$

where $A_{n,0}$ is the coefficient in the no heat generation case:

$$A_{n,0} = \dfrac{2Bi}{\psi_{\delta_n}[\delta_n^2 + Bi^2 - (\Gamma-1)Bi]} \tag{74}$$

Therefore, the complete temperature function becomes:

$$Y = \frac{T-T_{ex}}{\Delta T_0} = \frac{\beta}{\alpha^2}\left[\frac{Bi\cdot\psi(\alpha\cdot x)}{\alpha\cdot\psi'(\alpha)+Bi\cdot\psi(\alpha)} - 1\right] + \sum_{n=1}^{\infty} A_n\psi(\delta_n x)e^{-(\delta_n^2-\alpha^2)Fo} \tag{75}$$

## Average Value

The mass average temperature of the product at constant density is as follows (see [48]):

$$\bar{Y} = \bar{Y}_s + \sum_{n=1}^{\infty} \overline{A_n}e^{-(\delta_n^2-\alpha^2)Fo} \tag{76}$$

with:

$$\bar{Y}_s = -\frac{\beta}{\alpha^2}\left[\frac{(\Gamma+1)}{\alpha} \cdot \frac{Bi\cdot\psi'_\alpha}{\alpha\cdot\psi'_\alpha+Bi\cdot\psi_\alpha} + 1\right] \tag{77}$$

$$\overline{A_n} = \overline{A_{n,0}}\left(1 - \frac{\beta}{\delta_n^2-\alpha^2}\right) \tag{78}$$

and:

$$\overline{A_{n,0}} = A_{n,0}\overline{\psi_{\delta_n}} = \frac{2Bi^2(\Gamma+1)}{\delta_n^2[\delta_n^2+Bi^2-(\Gamma-1)Bi]} \tag{79}$$

Therefore, the mass average temperature function becomes:

$$\bar{Y} = \frac{\bar{T}-T_{ex}}{\Delta T_0} = -\frac{\beta}{\alpha^2}\left[\frac{(\Gamma+1)}{\alpha} \cdot \frac{Bi\cdot\psi'_\alpha}{\alpha\cdot\psi'_\alpha+Bi\cdot\psi_\alpha} + 1\right] + \sum_{n=1}^{\infty} \overline{A_n}e^{-(\delta_n^2-\alpha^2)Fo} \tag{80}$$

## Estimations and Applications

### *Cooling/Heating Times*

Similar to the case with no internal source, cooling times for sufficiently large Fourier numbers can be calculated using a very simple linear analytical expression. In fact, as in the case of chilling without an internal heat source, when the time is long enough the calculation can be done with the first term of the series. In this case Eq. (70) may be rewritten approximately as follows:

$$\theta(x, Fo) \equiv \theta = Y(x, Fo) - Y_s(x) \approx A \cdot \psi(\delta x) \cdot e^{-(\delta^2 - \alpha^2)Fo} \qquad (81)$$

Mass average temperature can be approximated just the same but substituting $A \cdot \psi(\delta x)$ for $\bar{A}$.

Fourier number (dimensionless cooling/heating time) can immediately be found solving for $Fo$ in Eq. (81). In fact:

$$\ln \theta = \ln(A\psi(\delta x)) - (\delta^2 - \alpha^2)Fo \qquad (82)$$

And solving for $Fo$:

$$Fo \approx \frac{\ln\left(\frac{A\psi}{\theta}\right)}{\delta^2 - \alpha^2} \qquad (83)$$

In the centre $\psi = 1$ and:

$$Fo_c \approx \frac{\ln\left(\frac{A}{\theta}\right)}{\delta^2 - \alpha^2} \qquad (84)$$

where the subscript $c$ denotes the value at the centre.

For the mass average value:

$$\overline{Fo} \approx \frac{\ln\left(\frac{A\bar{\psi}}{\theta}\right)}{\delta^2 - \alpha^2} \qquad (85)$$

**Displacement Correction**

From Eqs. (83) and (84) it can be written:

$$Fo \approx \frac{\ln\left(\frac{A\psi}{\theta}\right)}{\delta^2 - \alpha^2} = \frac{\ln\left(\frac{A}{\theta}\right)}{\delta^2 - \alpha^2} + \frac{\ln(\psi)}{\delta^2 - \alpha^2} \qquad (86)$$

That is:

$$Fo(\theta, x) \approx Fo_c(\theta) - D_x \qquad (87)$$

where:

$$D_x = \frac{\ln[1/\psi(\delta x)]}{\delta^2 - \alpha^2} \qquad (88)$$

Clearly this term is dependent not on time but solely on the coordinate. Hence, other than for low Fourier numbers, if we wish to calculate the time required at a coordinate x to reach a given temperature, $Y$, it will suffice to know the time needed to calculate it at the core and adjust this with the appropriate term for displacement. In the particular cases of the surface and the average value we get:

A) Surface:

$$D_{sf} = \frac{\ln(1/\psi_\delta)}{\delta^2 - \alpha^2} \qquad (89)$$

B) Average value:

$$\bar{D} = \frac{\ln(1/\bar{\psi})}{\delta^2 - \alpha^2} \qquad (90)$$

**Summary of the Procedure**

Hence, in the exponential zone (where the first term is enough), if we wish to know the time $Fo$ needed to attain the absolute dimensionless temperatures difference, $\theta$, it will suffice to apply the following procedure:

I. Take the difference value: $\theta = Y - Y_s$
II. Calculate the time for the core (Eq. 83)

III. Carry out the appropriate displacement if applicable [Eqs. (88), (89) or (90)]:

And finally import to Eq. (87).

## *Example IV (from Reference [48])*

The issue is to cool an individual potato using a 5°C flow of air, from an initial temperature of 25°C down to a temperature of 10°C in the centre. The surface heat transfer coefficient is $h = 3.0$ J/(m²s K).

According to Xu and Burfoot [47], we may assume that this potato is like a sphere 0.065 m in diameter. In this example, the model adopted by the authors just cited is also used. This is the linear model with temperature for the respiration heat:

$$q_r = aT + b$$

where $a = 6.99$ J/kg h K and $b = 62.62$ J/kg h.

Consequently, the parameters $A_0$ and $A_1$, as previously stated in SI units, are:

$$q_0 = b/3600 = 0.01739 \text{ J/(kg s)}$$

and

$$q_1 = a/3600 = 0.001942 \text{ J/(kg s K)}$$

Other values for $q_0$ and $q_1$, obtained from the linear regressions of average values taken from ASHRAE [12], can be seen in Table 1.

As the potato is assimilated to a sphere, the geometrical constant is $\Gamma = 2$.

The radius of the sphere is $R = 0.065/2 = 0.0325$ m

The parameters $q$ and $q_1$ in this chapter are measured in W/m³ and W/(m³K) respectively. Therefore, the previous values for $q$ and $q_1$ must be multiplied by the density. To do this and to calculate the Biot number the thermo-physical coefficients can be estimated from the chemical composition of the potato, the same as in the *Example II*.

The Table 3 in ASHRAE [12] (see pp.19.3) gives the following composition for potato:

$x_{Water} = 0.79$;   $x_{Prot} = 0.0207$;   $x_{fat} = 0.001$;   $x_{Carboh} 0.1798$; $x_{Ash} = 0.0089$

From these compositional values, following the same procedure as in the *Example II*, the following values are calculated for the thermo-physical parameters:

Density  $\rho = 1123.5 \text{ kg/m}^3$; thermal conductivity $k = 0.485$ W/m K; thermal diffusivity $a = 1.1877 \, 10^{-7} \text{ m}^2/s$; specific heat per unit mass $c_p = 3636.6 \text{ J/kg K}$.

Consequently, the respective values of $q_0$ and $q_1$ are finally:

$$q_0 = 0.01739 \times 1123.5 = 19.54 \text{ W/m}^3$$

And

$$q_1 = 0.001942 \times 1123.5 = 2.1816 \text{ W/(m}^3\text{K)}$$

And the Biot number is (Eq. (12)):

$$Bi = \frac{h \cdot R}{k} = \frac{3 \times 0.0325}{0.485} = 0.201$$

The dimensionless parameters $\alpha^2$ and $\beta$ are (Eqs. (66)-(68)):

$$\alpha^2 = \frac{q_1 R^2}{k} = \frac{2.1816 \times 0.0325^2}{0.485} = 0.00475$$

Whose value depends solely on the specific fruit and not on the process.

The external temperature is 5°C and the total temperature difference is:

$$\Delta T_0 = T_0 - T_{ex} = 25 - 5 = 20°C$$

Therefore, the value of $\beta$ is:

$$\beta = \frac{B_0 R^2}{k \Delta T_0} = \frac{(q_0 + q_1 T_{ex}) R^2}{k(T_0 - T_{ex})} = \frac{(19.54 + 2.1816 \times 5) \times 0.0325^2}{0.485 \times 20} = 0.00331$$

This depends on both the potato itself and the cooling/heating process, because the external temperature and the total temperature difference affect it.

As the potato is considered as a sphere, the value of $\delta$ is $\delta = 0.7610$ (and hence $\delta^2 = 0.5791$), as it can be proved by substituting this value into the transcendental boundary equation (Eq. (72)) applied to the sphere:

$$\delta \cos \delta - (1 - Bi) * \sin \delta = 0$$

From Eq. (74) the $A$ value for the no heat of respiration is:

$$A_0 = \frac{2Bi}{\psi_{\delta_n}[\delta_n^2 + Bi^2 - (\Gamma - 1)Bi]} = \frac{2 \times 0.201}{\frac{\sin 0.7610}{0.7610}[0.5791 + 0.201^2 - (2-1)0.201]} = 1.0594$$

Accordingly, introducing this value into Eq. (73):

$$A = A_0 \left(1 - \frac{\beta}{\delta_n^2 - \alpha^2}\right) = 1.0594 \left(1 - \frac{0.00331}{0.5791 - 0.00475}\right) = 1.0533$$

The steady temperature at the core is (Eq. (71)) (remember that at $x = 0$ $\psi = 1$):

$$Y_s(x) = \frac{T_s - T_{ex}}{\Delta T_0} = \frac{\beta}{\alpha^2}\left[\frac{Bi}{\alpha \cdot \psi'(\alpha) + Bi \cdot \psi(\alpha)} - 1\right] = 0.0061$$

The temperature to be reach at the centre is $T_c = 10\ °C$

$$Y_c = \frac{10-5}{20} = 0.25$$

Consequently (Eq. (81)):

$$\theta_c = Y_c - Y_s = 0.25 - 0.0061 = 0.2439$$

And (Eq. (84)):

$$Fo_c \approx \frac{\ln\left(\frac{A}{\theta}\right)}{\delta^2 - \alpha^2} = \frac{\ln\left(\frac{1.0533}{0.2439}\right)}{0.5791 - 0.00475} = 2.547$$

Which coincides with the result from the complete series.
Finally, clearing $t$ in (Eq. (11)):

$$Fo = \frac{\alpha \cdot t}{R^2}$$

$$t_c = \frac{Fo_c R^2}{a} = \frac{2.547 \times 0.0325^2}{1.1877 \times 10^{-7}} = 22652\ s = 6.29\ h$$

At this moment, what is the mass average temperature?

As mentioned above, Eq. (81) can be applied but substituting $Y_s$ for $\bar{Y}_s$ (Eq. (77)) and $A$ for $\bar{A}$ (Eqs. (78) and (79)). The results are:

$$\bar{Y}_s = 0.0058$$
$$\bar{A} = 0.9936$$

And finally:

## Cooling Kinetics in Stone Fruits

$$\bar{Y} \approx \bar{Y}_s + \bar{A} \cdot e^{-(\delta^2 - \alpha^2)Fo} = 0.0058 + 0.9936 \times e^{-(0.5791 - 0.00475) \times 2.547}$$
$$= 0.2358$$

That is (remember that $\bar{Y} = (\bar{T} - T_{ex})/\Delta T_0$)

$$\bar{T} = T_{ex} + \Delta T_0 \bar{Y} = 5 + 20 \times 0.2358 = 9.72\ °C$$

The values for this example in the case where the heat of respiration is not considered are:

Core: $Fo = 2.494$ (difference -2.10%)

Mass average temperature: $\bar{T} = 9.57\ °C$ (difference -1.48%)

### Maximum Value at the Core

Figure 5 depicts the temperature history for this example calculated with the sum of the first 200 addends of the complete Fourier series and the first 53 seconds (approximately: up to $Fo = 0.06$).

As it can be seen in the Figure 5, the temperature rises at the outset up to a given maximum value $Y_M$ in a certain time, $Fo_M$, falling thereafter down to its steady value.

At this point $\{Fo_M, Y_M\}$ the derivative of $Y$ with respect to $Fo$ should be zero. Thus, deriving with respect to $Fo$ in Eq. (70) (at the centre):

$$\left.\frac{dY}{dFo}\right|_{Fo_M} = -\sum_{n=1}^{\infty}(\delta_n^2 - \alpha^2)A_n e^{-(\delta_n^2 - \alpha^2)Fo} = 0 \tag{91}$$

Taking into account Eq. (73):

$$A_n(\delta_n^2 - \alpha^2) = A_{n,0}(\delta_n^2 - \alpha^2)\left(1 - \frac{\beta}{\delta_n^2 - \alpha^2}\right) = A_{n,0}(\delta_n^2 - \alpha^2 - \beta) \tag{92}$$

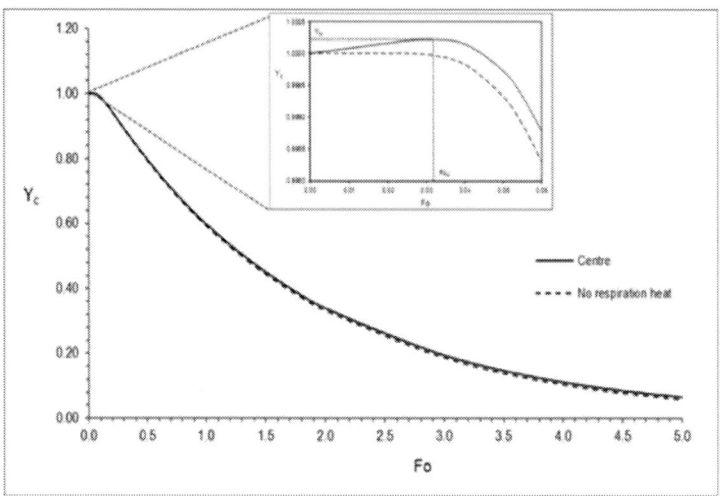

Figure 5. Temperature history for the *Example IV*, depicted with the sum of the first 200 addends of the complete Fourier series.

Consequently, introducing Eq. (92) into Eq. (91) and rearranging terms, it follows that, at $\{Fo_M, Y_M\}$ the following condition must be met:

$$\sum_{n=1}^{\infty} A_{n,0} \delta_n^2 e^{-(\delta_n^2 - \alpha^2) Fo_M} = (\alpha^2 + \beta) \sum_{n=1}^{\infty} A_{n,0} e^{-(\delta_n^2 - \alpha^2) Fo_M}$$

That is:

$$\frac{\sum_{n=1}^{\infty} A_{n,0} \delta_n^2 e^{-(\delta_n^2 - \alpha^2) Fo_M}}{\sum_{n=1}^{\infty} A_{n,0} e^{-(\delta_n^2 - \alpha^2) Fo_M}} = (\alpha^2 + \beta)$$

But whatever:

$$e^{-(\delta_n^2 - \alpha^2) Fo_M} = e^{-\delta_n^2 Fo_M} e^{\alpha^2 Fo_M}$$

$$\frac{\sum_{n=1}^{\infty} A_{n,0} \delta_n^2 e^{-\delta_n^2 Fo_M} e^{\alpha^2 Fo_M}}{\sum_{n=1}^{\infty} A_{n,0} e^{-\delta_n^2 Fo_M} e^{\alpha^2 Fo_M}} = (\alpha^2 + \beta)$$

As $Fo_M$ must have a concrete value, $e^{\alpha^2 Fo_M}$ is a common factor that can be extracted from both sums, cancelling itself between the numerator and the denominator, leaving:

$$\frac{\sum_{n=1}^{\infty} A_{n,0} \delta_n^2 e^{-\delta_n^2 Fo_M}}{\sum_{n=1}^{\infty} A_{n,0} e^{-\delta_n^2 Fo_M}} = (\alpha^2 + \beta) \qquad (93)$$

Eq. (93) can also be written as:

$$-\frac{\left.\frac{dY_0}{dFo}\right|_{Fo_M}}{Y_0(Fo_M)} = (\alpha^2 + \beta) \qquad (94)$$

Where $Y_0$ is the dimensionless temperature difference for the no respiration case corresponding to the same Biot number.

If we substitute this value $Fo_M$ in Eq. (70):

$$Y_M = Y_{s,c} + \sum_{n=1}^{\infty} A_n e^{-(\delta_n^2 - \alpha^2) Fo_M} \qquad (95)$$

In the *Example IV*, Eq. (95) holds for $Fo_M = 0.0318$, $Y_M = 1.0002$. Figure 5 also depicts the first moments in the complete curve, when the maximum point $\{Fo_M, Y_M\}$ is reached.

That is, in this particular case the influence of the respiration heat is practically negligible, because the values of $q_0$ and $q_1$ are low, the mass considered is very small (an individual potato) and (as it will be seen bellow) the Biot number is much larger than the threshold value, the influence of the respiration heat is relatively small. However, as can be seen in Figure 6, for much higher values the influence can be significant.

Figures 6 and 7 illustrate the influence of the parameters $\alpha^2$ and $\beta$ by plotting the temperature history for a fixed Biot number ($Bi = 5$) and for different values of $\alpha^2$ and $\beta$. In Figure 6, the value of $\beta$ ($\beta = 1$) is fixed and that of $\alpha^2$ varies between $\alpha^2 = 1$ and $\alpha^2 = 5$ (always lower than the first value of $\delta^2$ corresponding to the Biot number). However, in Figure 7 $\alpha^2$ is fixed ($\alpha^2 = 1$) and $\beta$ varies between $\beta = 1$ and $\beta = 5$.

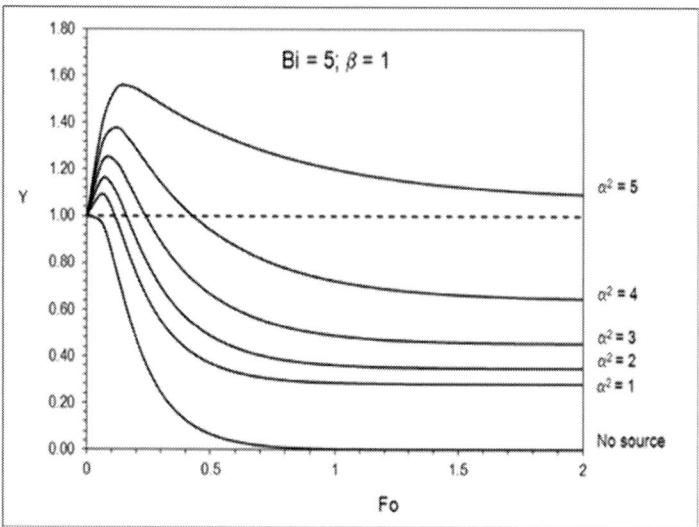

Figure 6. Temperature history for $\beta = 1 = Cte$ and different values of $\alpha^2$.

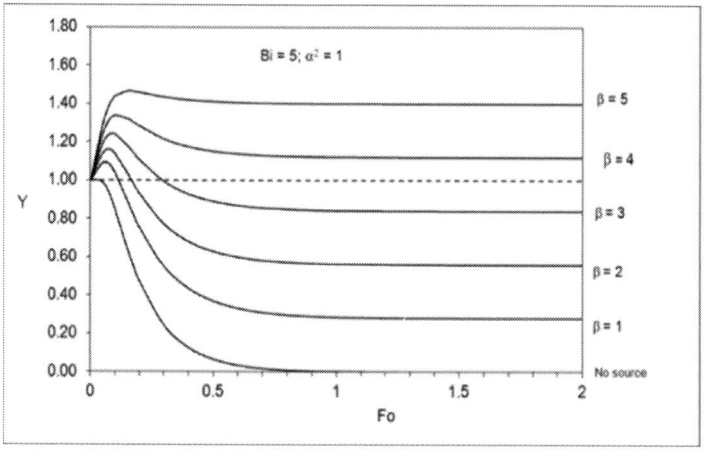

Figure 7. Temperature history for $\alpha^2 = 1 = Cte$ and different values of $\beta$.

## Threshold Biot Number

As shown by Cuesta and Lamúa [48], $\alpha^2$ must satisfy the condition:

$$\alpha^2 \leq \delta_1^{\,2} \leq \delta_{1,M}^{\,2} \tag{96}$$

However, Eq. (93) also means an absolute lower limit of the Biot number for the fruit to be able to cool. In fact, Eq. (93) determines the time to get the maximum value of the temperature. For $Fo \to \infty$, the quotient $-(dY/dFo)/Y \to \delta_1^2$. In fact, for $Fo \to \infty$, only the first addend is significant in both the numerator and the denominator:

$$\frac{\sum_{n=1}^{\infty} A_{n,0} \delta_n^2 e^{-\delta_n^2 Fo}}{\sum_{n=1}^{\infty} A_{n,0} e^{-\delta_n^2 Fo}} \to \frac{A_{1,0} \delta_1^2 e^{-\delta_1^2 Fo}}{A_{1,0} e^{-\delta_1^2 Fo}} = \delta_1^2 \tag{97}$$

Consequently, $Fo_M \to \infty$ means that the fruit is always heating as a consequence of the internal generation of heat. From Eqs. (94) and (97), when $Fo_M \to \infty$:

$$-\frac{\left.\frac{dY_0}{dFo}\right|_{Fo_M}}{Y_0(Fo_M)} = \delta_1^2 = (\alpha^2 + \beta) \tag{98}$$

That is, so that the fruit can cool down, the Biot number must be such that the first eigenvalue ($\delta_1$) of boundary Eq. (72) must meet the condition:

$$\delta_1^2 > (\alpha^2 + \beta) \tag{99}$$

Therefore, the threshold Biot number ($Bi_{Th}$) is the one whose first eigenvalue of boundary Eq. (72) is:

$$\delta_{1,Th}^2 = (\alpha^2 + \beta) \tag{100}$$

As $\delta_{1,Th}$ must also be an eigenvalue of Eq. (72):

$$\delta_{Th} \cdot \psi'_{\delta_{Th}} = -Bi_{Th} \cdot \psi_{\delta_{Th}}$$

And, solving for $Bi_{Th}$:

$$Bi_{Th} = -\delta_{Th} \cdot \frac{\psi'_{\delta_{Th}}}{\psi_{\delta_{Th}}} \tag{101}$$

Remember that for the flat plate $\psi_{\delta_{Th}} = \cos\delta_{Th}$ and $\left(-\psi'_{\delta_{Th}}/\psi_{\delta_{Th}}\right) = \tan\delta_{Th}$, for the infinite cylinder $\psi_{\delta_{Th}} = J_0(\delta_{Th})$ and $\left(-\psi'_{\delta_{Th}}/\psi_{\delta_{Th}}\right) = J_1(\delta_{Th})/J_0(\delta_{Th})$, and for the sphere $\psi_{\delta_{Th}} = \sin\delta_{Th}/\delta_{Th}$ and $\left(-\psi'_{\delta_{Th}}/\psi_{\delta_{Th}}\right) = 1/\delta_{Th} - \cot(\delta_{Th})$.

Below, in Figure (13), Eq. (101) is represented for solid sphere.

According to Cuesta et al. [23], and taking into account Eq. (100), the threshold Biot number can also be calculated from the exact numerical series:

$$Bi_{Th} = \frac{h_{Th} \cdot R}{k} = (\alpha^2 + \beta)\left[\frac{1}{(\Gamma+1)} + \sum_1^\infty \frac{2}{(\delta_{n,M})^2} \cdot \frac{(\alpha^2+\beta)}{(\delta_{n,M})^2 - (\alpha^2+\beta)}\right] \tag{102}$$

$(\Gamma + 1)$ is the geometric constant (1 for an infinite slab, 2 for an infinite cylinder and 3 for a sphere) and $\delta_{n,M}$ are the values of $\delta_n$ when $Bi \to \infty$: $\delta_{n,M} = (2n-1)\pi/2$ for an infinite slab; $\delta_{n,M} = n \cdot \pi$ for a sphere. In the case of an infinite cylinder, the values have been tabulated [49].

### *Estimation to $Fo_M$, $Y_M$ and $Bi_{Th}$*

There are two ways to get a first approximation to $Fo_M$, $Y_M$ and $Bi_{Th}$ with which to start the iterative trial error process to solve the Eqs. (94) and (65).

The firs one is based on the nomogram represented in Figure 8. Figure 8a depicts $(\alpha^2 + \beta)$ vs. $Fo_M$ for different Biot numbers. Actually the depicted curves are the quotient $-\left(\frac{dY_0}{dFo}\right)/Y_0$ vs. $Fo$, for different Biot numbers, although by virtue of Eq. (94), when choosing a particular value as $(\alpha^2 + \beta)$, the corresponding $Fo_M$ value is also determined. For this reason, the ordinate represents the sum $(\alpha^2 + \beta)$ and the abscissa $Fo_M$ (at the top of the Figure 8a). Figure 8b depicts the first moments in the temperature kinetic for the same Biot numbers as shown in Figure 8a and for $(\alpha^2 + \beta) = 1$, to illustrate the example. Figure 8c depicts $(\alpha^2 + \beta)$ vs. $Bi_{Th}$. As in the Figure 8a, the curve represented here actually depicts the square of the first eigenvalue of Eq. (72) for the sphere ($\delta_1^2$ vs. $Bi$) but, by virtue of Eq. (100), when choosing as particular value $\delta_1^2 = (\alpha^2 + \beta)$, the corresponding $Bi_{Th}$ value is determined.

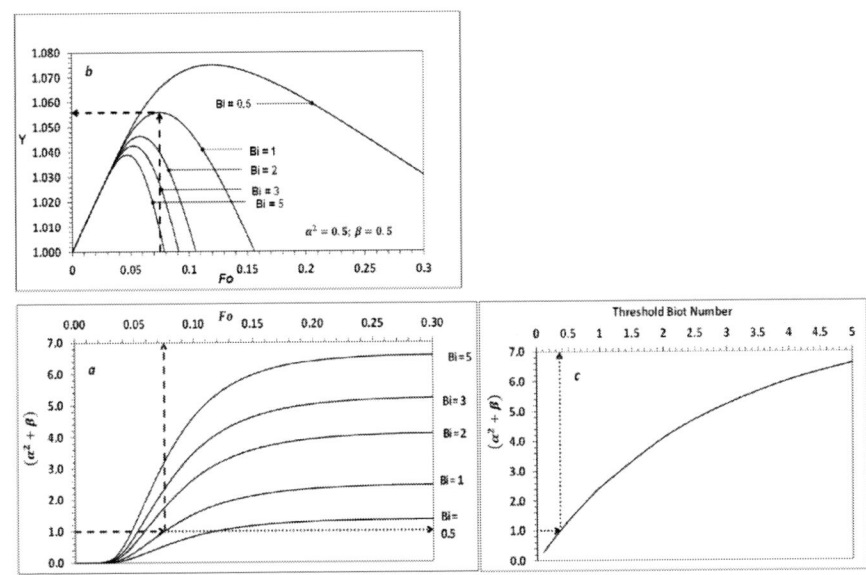

Figure 8. (a): $(\alpha^2 + \beta)$ vs. $Fo_M$ for different Biot numbers. (b): Beginning of the $\{Y - Fo\}$ curve for the same Biot numbers as in (a) and $(\alpha^2 + \beta) = 1$. (c) $(\alpha^2 + \beta)$ vs. $Bi_{Th}$.

As an example, the straight line parallel to the abscissa in Figure 8(a) for the particular value $(\alpha^2 + \beta) = 1$ intersects with the curve corresponding to the desired Biot number (in this case $Bi = 1$), and the vertical line from this point to the abscissa (at the top of Figure 8a) gives us the corresponding value of $Fo_M$ (in this example $Fo_M = 0.075$). This dashed line can be elongated to the same Biot number in the Figure 8(b) to determine the maximum $Y$ ($Y_M$) corresponding to $Fo_M$ (in this example

$Y_M = 1.056$). The horizontal dashed line in part a of the figure can also be extended parallel to abscissa (dotted line) to enter in the Figure 8(c) to determine the corresponding threshold Biot number (in this case $Bi_{Th} = 0.358$).

On the other hand, the second way to get a first approximation to $Fo_M$ for very low Biot numbers is to consider only the two first addends in Eq. (94). The result is:

$$Fo_M \approx \frac{\ln\left[-\frac{A_{2,0}\left[\delta_2^2-(\alpha^2+\beta)\right]}{A_{1,0}\left[\delta_1^2-(\alpha^2+\beta)\right]}\right]}{(\delta_2^2-\delta_1^2)} \quad (103)$$

The smaller $\alpha^2$ is, the more imprecise Eq. (103) is, because for $\alpha^2 = 0$ (no heat of respiration case) the null derivative is satisfied only at the initial instant.

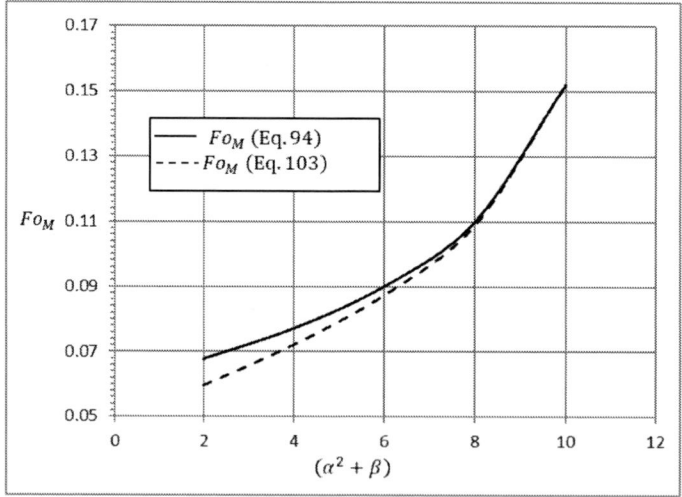

Figure 9. Comparison between the first approximation to $Fo_M$ (Eq. (103)) and the actual value $Fo_M$ obtained with the complete series (Eq. (94)).

For the case described in the *Example IV*, by substituting $\delta_1 = 0.7610$; $\delta_2 = 4.538$; $A_1 = 1.0533$; $A_2 = -0.0906$ and $\alpha^2 = 0.00475$ into Eq. (103), the result is $Fo_M \approx 0.0563$ ($\approx 500 \text{ s} = 8$ min), very far from the actual value seen in the *Example IV* ($Fo_M = 0.0318$) because ($\alpha^2 + \beta$) is very close to the no respiration case.

Figure 9 depicts the comparison between the first approximation to $Fo_M$ (Eq. (103)) and the actual value $Fo_M$ obtained with the complete series (Eq. (94)).

According to Cuesta et al. [23], to estimate the threshold Biot number, next equation can be used:

$$Bi_{Th} \approx (\alpha^2 + \beta)\left[\frac{1}{(\Gamma+1)} + \frac{2}{(\delta_{1,M})^2} \cdot \frac{\alpha^2+\beta}{(\delta_{1,M})^2-\alpha^2-\beta}\right] \quad (104)$$

This equation is valid for the three elementary geometries and shows the Biot number as function of the slope $\delta_1^2$ and the particular geometry of the object (constants $\Gamma$ and $\delta_{1,M}$).

For the particular case of the sphere $(\Gamma + 1) = 3$ and $\delta_{1,M} = \pi$:

$$Bi_{Th} \approx (\alpha^2 + \beta)\left[\frac{1}{3} + \frac{2}{\pi^2} \cdot \frac{(\alpha^2+\beta)}{\pi^2-(\alpha^2+\beta)}\right] \quad (105)$$

In the example of the Figure 8, the threshold Biot number deduced from Eq. (105) is $Bi_{Th} \approx 0.356$ (that is, the deviation is -0.48%). In the case of the *Example IV*, $(\alpha^2 + \beta) = 0.008062$, and introducing this value into Eq. (105) the result is $Bi_{Th} \approx 0.00269$. The exact is $Bi_{Th} = 0.00269$ (that is, the deviation is -0.03%), much lower than the Biot number in the same example.

# MODELLING THERMAL KINETICS IN STONE FRUITS CONSIDERING HEAT OF RESPIRATION LINEARLY RELIANT ON TEMPERATURE

As previously said, in this section the third modification of the Fourier equations is introduced (the joint modification produced by both at the same time), which consists of considering the internal source of heat produced by the respiration into a stone fruit.

Thus, this section is devoted to model the thermal kinetic in stone fruits, but also considering the internal heat of respiration as a linear function on temperature.

## Mathematical Background

Suppose, as in the second section, a fruit whose shape is considered spherical with outer radius $R$ and whose seed is considered a sphere concentric with the previous one and radius $r_0 < R$ (Figure 1). This core is considered quasi-insulating, for purposes of heat conduction, compared to the external heat transfer in the times that the process lasts. It is assumed that the initial temperature of the sphere is uniform ($T_0$), and that the fruit is placed in sudden contact with a mechanically-stirred medium at a uniform temperature $T_{ex}$. The fruit is considered to have an internal heat source linearly dependent on temperature. Thus, the main governing equation is Eq. (63):

$$k \frac{1}{r^2} \frac{\partial}{\partial r}\left(r^2 \frac{\partial T}{\partial r}\right) + q_0 + q_1 T = \rho c \frac{\partial T}{\partial t}$$

With the same external boundary condition:

$r = R$:
$$k \left[\frac{\partial T}{\partial r}\right]_{r=R} = -h(T_{Sf} - T_{ex})$$

$T_{Sf}$ being the temperature of the pulp at the external surface.

But with the internal boundary condition at the stone–pulp interface as in second section, Eq. (27):

$$\left[\frac{\partial T}{\partial r}\right]_{r=r_0} = 0$$

As seen in the previous sections, the above equations may be written in dimensionless form thus:

$$\frac{1}{x^2}\frac{\partial}{\partial x}\left(x^2\frac{\partial \theta}{\partial x}\right) + \alpha^2\theta + \beta = \frac{\partial \theta}{\partial Fo} \tag{106}$$

$$\left[\frac{\partial Y}{\partial x}\right]_{x=1} = -Bi \cdot Y_{Sf} \tag{107}$$

$$\left[\frac{\partial Y}{\partial x}\right]_{x=x_0} = 0 \tag{108}$$

As seen in the above second section (Modelling Thermal Kinetics in Stone Fruits), and according to Cuesta and Lamúa [11] and Cuesta and Alvarez [14], the heat source is expressed by two dimensionless numbers:

$$\alpha^2 = \frac{q_1 R^2}{k} \tag{109}$$

$$\beta = \frac{B_0 R^2}{k \cdot \Delta T_0} \tag{110}$$

where $B_0$ is:

$$B_0 = q_0 + q_1 T_{ex} \tag{111}$$

The general solution to Eq. (106) with the external and internal stone–pulp interface boundary conditions (107) and (108), respectively is in accordance with Cuesta and Alvarez [14]:

$$Y(x, Fo) = Y_s(x) + \sum_{i=1}^{\infty} A_i \cdot \psi(\delta_i x) \cdot e^{-(\delta_i^2 - \alpha^2) Fo} \tag{112}$$

where:

$$\psi(\delta_i x) = \frac{\sin[\delta_i(x-x_0)] + \delta_i x_0 \cos[\delta_i(x-x_0)]}{\delta x} \tag{113}$$

And $Y_s$ is the steady function:

$$Y_s(\alpha x) = \frac{\beta}{\alpha^2}\left[\frac{Bi\cdot\psi_s(\alpha x)}{\alpha\psi'_{s,\alpha}+Bi\cdot\psi_{s,\alpha}} - 1\right] \tag{114}$$

Eq. (114) is formally the same as Eq. (70), except for $\psi_s(\alpha x)$:

$$\psi_s(\alpha x) = \frac{\sin[\alpha(x-x_0)]}{\alpha\cdot x} + \frac{x_0\cdot\cos[\alpha(x-x_0)]}{x} \tag{115}$$

And constants $\delta_i$, which are the eigenvalues of the boundary [14]:

$$\delta = \tan[\delta(1-x_0)]\frac{1-Bi+\delta^2 x_0}{1+x_0(Bi-1)} \tag{116}$$

And the expansion constants $A_i$ are [14]:

$$A_i = A_{i,0}\left(1 - \frac{\beta}{\delta_i^2 - \alpha^2}\right) \tag{117}$$

with

$$A_{i,0} = \frac{2Bi\psi(\delta_i)}{[\psi(\delta_i)]^2(\delta_i^2 + Bi^2 - Bi) - \delta_i^2 x_0^3} \tag{118}$$

As can be seen, constants $A_i$ (Eq. 118) also are formally identical to those obtained in the third section (Modelling Thermal Kinetics Considering Internal Linearly Temperature Dependent Heat Generation) (Eq. (73)), but constants $\delta_i$ and $A_{i,0}$, as well as the functions $\psi(\delta_i x)$ and $\psi_s(\alpha x)$, are calculated according to the second section (Modelling Thermal Kinetics in Stone Fruits, Eqs. (31), (32) and (33)).

In order to illustrate, the Figure 10 shows the influence of the parameter $\alpha^2$ by plotting the temperature histories at the thermal centre (stone–pulp contact) for a fixed $x_0$ ($x_0 = 0.5$), Biot number ($Bi = 5$) and $\beta$ ($\beta = 1$), and for different values of $\alpha^2$ ($\alpha^2 = 1$ to 5). The Figure 11 shows the influence of the parameter $\beta$ by plotting the temperature histories at the thermal centre (stone–pulp contact) for a fixed $x_0$ ($x_0 = 0.5$), Biot number ($Bi = 5$) and $\alpha^2$ ($\alpha^2 = 1$), and for different values of $\beta$ ($\beta = 1$ to $5$). In turn, the Figure 12 shows the influence of the parameter $x_0$ by plotting the temperature histories at the thermal centre (stone–pulp contact) for a fixed Biot number ($Bi = 5$), $\alpha^2$ ($\alpha^2 = 2$) and $\beta$ ($\beta = 3$), and for different values of $x_0$ ($x_0 = 0.3$ to $0.6$).

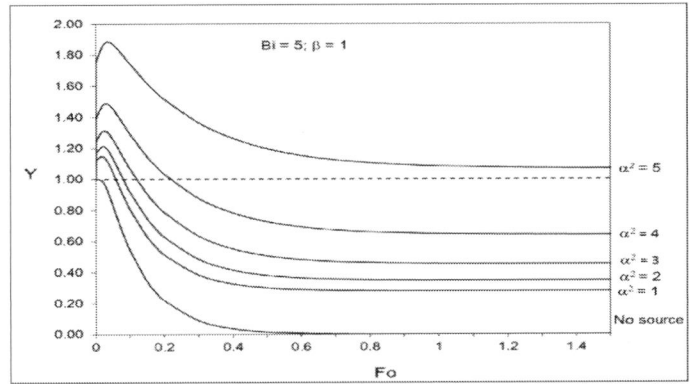

Figure 10. Temperature histories for $x_0 = 0.5$, $Bi = 5$, $\beta = 1$ and different values of $\alpha^2$. The crosses represent the maximum values ($Y_M$) corresponding to the values $Fo_M$ obtained by Eq. (124).

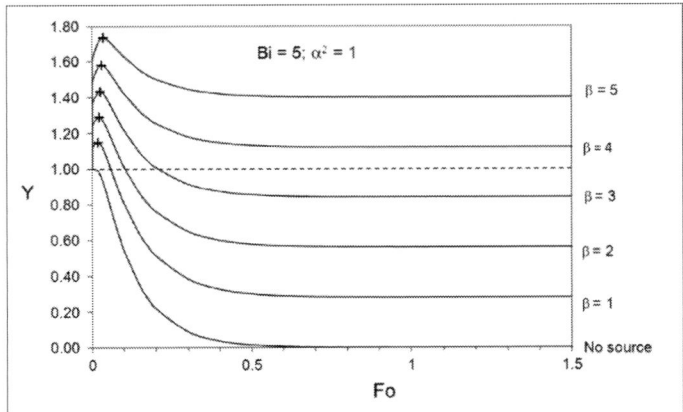

Figure 11. Temperature histories for $x_0 = 0.5$, $Bi = 5$, $\alpha^2 = 1$ and different values of $\beta$. The crosses represent the maximum values ($Y_M$) corresponding to the values $Fo_M$ obtained by Eq. (124).

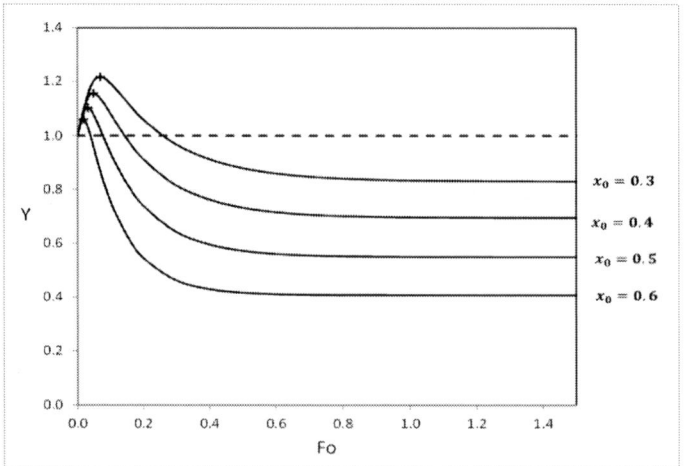

Figure 12. Temperature histories for $Bi = 5$, $\alpha^2 = 2$, $\beta = 3$ and different values of $x_0$. The crosses represent the maximum values $(Y_M)$ corresponding to the values $Fo_M$ obtained by Eq. (124).

As the density is considered to be constant, the mass average temperature is:

$$\bar{Y}(Fo) \equiv \bar{Y} = \frac{1}{V}\int_{x_0}^{1} Y(x, Fo) dV = \bar{Y}_s + \sum_{i=1}^{\infty} \bar{A}_i \cdot e^{-(\delta_i^2 - \alpha^2)Fo} \quad (119)$$

where, as shown in Cuesta and Alvarez [13]:

$$\bar{Y}_s = -\frac{\beta}{\alpha^2}\left[\frac{3}{\alpha\cdot(1-x_0^3)} \cdot \frac{Bi\cdot\psi'_{s,\alpha}}{\alpha\psi'_{s,\alpha} + Bi\cdot\psi_{s,\alpha}} + 1\right] \quad (120)$$

and:

$$\bar{Y}_i = \frac{6Bi^2}{\delta_i^2(1-x_0^3)} \frac{\psi^2(\delta_i)}{[\psi(\delta_i)]^2(\delta_i^2 + Bi^2 - Bi) - \delta_i^2 x_0^3}\left(1 - \frac{\beta}{\delta_i^2 - \alpha^2}\right) \quad (121)$$

And when $Bi \to \infty$ (Eq. (120)) remains:

$$\overline{Y}_{s,\infty} = -\frac{\beta}{\alpha^2}\left[\frac{3}{\alpha \cdot (1-x_0^3)} \cdot \frac{\psi'_{s,\alpha}}{\psi_{s,\alpha}} + 1\right] \quad (122)$$

and

$$\overline{Y}_{i,\infty} = \frac{6}{\delta_{i,M}^2(1-x_0^3)} \cdot \frac{1+\delta_{i,M}^2 x_0^2}{1+\delta_{i,M}^2 x_0^2 - \delta_{i,M}^2 x_0^3} \cdot \left(1 - \frac{\beta}{\delta_{i,M}^2 - \alpha^2}\right) \quad (123)$$

## Maximum Value at the Core

As it can be seen in the Figures 10-12, Eq. (112) also reaches a maximum $Y_M$ in a certain time, $Fo_M$, falling thereafter down to its steady value. At this point:

$$\left.\frac{dY}{dFo}\right]_{Fo_M} = -\sum_{n=1}^{\infty}(\delta_n^2 - \alpha^2)A_n e^{-(\delta_n^2-\alpha^2)Fo} = 0$$

And rearranging terms like seen above in Eqs. (93) and (94), the result formally is the same:

$$\frac{\sum_{n=1}^{\infty} A_{n,0}\delta_n^2 e^{-\delta_n^2 Fo_M}}{\sum_{n=1}^{\infty} A_{n,0} e^{-\delta_n^2 Fo_M}} = (\alpha^2 + \beta) \quad (124)$$

$$-\frac{\left.\frac{dY_0}{dFo}\right]_{Fo_M}}{Y_0(Fo_M)} = (\alpha^2 + \beta) \quad (125)$$

where $Y_0$ is the dimensionless temperature difference for the no respiration case corresponding to the same Biot number.

And as shown in the third section (Modelling Thermal Kinetics Considering Internal Linearly Temperature Dependent Heat Generation), to cool the product, the rate of extraction of heat via the surface must be greater than the heat generated in the interior. This can be expressed by the same conditions (Eq. (96)):

$$\alpha^2 \leq \delta_1{}^2 \leq \delta_{1,M}{}^2$$

And as in Eqs. (97) and (98) for $Fo \to \infty$, only the first addend is significant in both the numerator and the denominator, and:

$$\frac{\sum_{n=1}^{\infty} A_{n,0}\delta_n{}^2 e^{-\delta_n{}^2 Fo}}{\sum_{n=1}^{\infty} A_{n,0} e^{-\delta_n{}^2 Fo}} \to \frac{A_{1,0}\delta_1{}^2 e^{-\delta_1{}^2 Fo}}{A_{1,0} e^{-\delta_1{}^2 Fo}} = \delta_1{}^2 \qquad (126)$$

$$-\frac{\left.\frac{dY_0}{dFo}\right|_{Fo_M}}{Y_0(Fo_M)} = \delta_1{}^2 = (\alpha^2 + \beta) \qquad (127)$$

The difference with Eqs. (93) and (94), and Eqs. (97) and (98), is that Eqs. (126) and (127) are explicit functins of $x_a$ (Figure 2).

## Threshold Biot Number

Consequently with Eqs. (126) and (127), the threshold Biot number can be written with the condition:

$$\delta_{1,Th}{}^2 = (\alpha^2 + \beta) \qquad (128)$$

where $\delta_{1,Th}$ must be an eigenvalue of Eq. (116):

$$\delta_{1,Th} = \tan[\delta_{1,Th}(1-x_0)] \frac{1 - Bi_{Th} + \delta_{1,Th}{}^2 x_0}{1 + x_0(Bi_{Th} - 1)} \qquad (129)$$

Consequently, solving for $Bi_{Th}$ in Eq. (129):

$$Bi_{Th} = 1 + \frac{(\delta_{1,Th})^2 x_0 - m}{1 + mx_0} \qquad (130)$$

where $m$ is:

$$m \equiv \frac{\delta_{1,Th}}{\tan[\delta_{1,Th}(1-x_0)]} \quad (131)$$

And, by substituting Eq. (128) into Eqs. (131) and (132):

$$Bi_{Th} = 1 + \frac{(\alpha^2+\beta)x_0 - m}{1+mx_0} \quad (132)$$

Where now:

$$m \equiv \frac{\sqrt{(\alpha^2+\beta)}}{\tan[(1-x_0)\sqrt{(\alpha^2+\beta)}]} \quad (133)$$

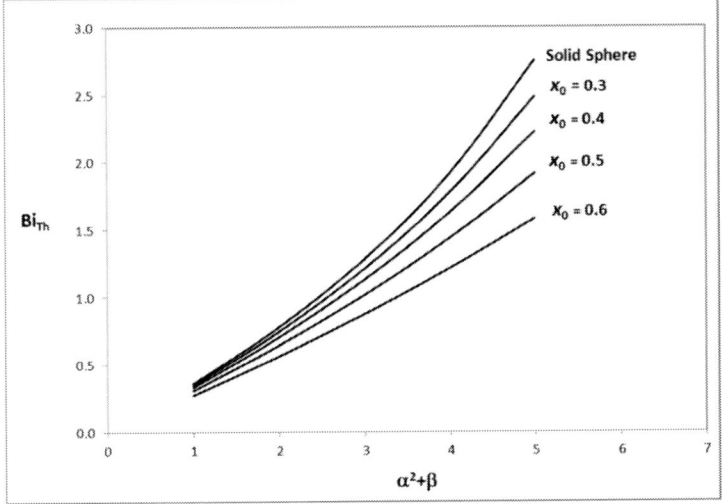

Figure 13. Threshold Biot number for $x_0 = 0$ (solid sphere) to $x_0 = 0.6$.

Figure (13) shows the threshold Biot number vs. $(\alpha^2 + \beta)$ in stone fruits for different values of $x_0$

If we substitute this value $Fo_M$ in Eq. (112):

$$Y_M = Y_{s,c} + \sum_{n=1}^{\infty} A_n e^{-(\delta_n^2 - \alpha^2)Fo_M} \quad (134)$$

The Figures 10–12 also show the maximum values of the dimensionless temperature $Y_M$ obtained from Eqs. (125) and (134).

## Estimations and Applications

### *Cooling/Heating Times*

As mentioned in all the previous cases, for practical purposes and from a given moment onwards, the exact solution in infinite series reduces very quickly to its first exponential term, whose representation in semi-logarithmic scale is a straight line. Thus, from Eq. (112):

$$\theta = Y(x, Fo) - Y_s(x) \approx A \cdot \psi(\delta x) \cdot e^{-(\delta^2 - \alpha^2)Fo} \qquad (135)$$

$$\ln \theta = \ln[A\psi(\delta x)] - (\delta^2 - \alpha^2)Fo \qquad (136)$$

$$Fo \approx \frac{\ln\left(\frac{A\psi}{\theta}\right)}{\delta^2 - \alpha^2} \qquad (137)$$

$$Fo_{x_0} \approx \frac{\ln\left(\frac{A}{\theta}\right)}{\delta^2 - \alpha^2} \qquad (138)$$

$$\overline{Fo} \approx \frac{\ln\left(\frac{A\bar{\psi}}{\theta}\right)}{\delta^2 - \alpha^2} \qquad (139)$$

### **Displacement Correction**

At the dimensionless distance $x = r/R$ clearing $Fo$ in Eq. (135):

$$Fo(x) \approx \frac{\ln A + \ln \psi(\delta \cdot x) - \ln Y}{(\delta^2 - \alpha^2)} \qquad (140)$$

Thus, to get the same temperature at $x_0$ and at the coordinate $x$, $Fo(x)$ can be rewritten as follows:

$$Fo(x) \approx \frac{\ln A/Y}{(\delta^2-\alpha^2)} + \frac{\ln \psi(\delta \cdot x)}{(\delta^2-\alpha^2)} = Fo_c - \frac{\ln(1/\psi(\delta \cdot x))}{(\delta^2-\alpha^2)}$$

That is, the correction on time for the coordinate $x$ is:

$$D_x \equiv Fo_c - Fo_x \approx \frac{\ln(1/\psi(\delta \cdot x))}{(\delta^2-\alpha^2)} \quad (141)$$

Clearly Eq. (141) is dependent not on the time, but solely on the coordinate. Hence, other than for low Fourier numbers, if we wish to calculate the time required at a coordinate $x$ to reach a given temperature $T$, it will be sufficient to know the time needed to calculate it at the core and adjust it with the appropriate term for displacement.

In order to know the temperature at the coordinate $x$ when is known at the stone-to-pulp surface $x_0$ (thermal centre):

At $x_0$:

$$Y_c - Y_{s,c} \approx A \cdot e^{-(\delta^2-\alpha^2)Fo} \quad (142)$$

At coordinate $x$:

$$Y_x - Y_{s,x} \approx A \cdot \psi(\delta \cdot x) \cdot e^{-(\delta^2-\alpha^2)Fo} = \psi(\delta \cdot x) \cdot (Y_c - Y_{s,c}) \quad (143)$$

That is, the correction on temperature for the coordinate $x$ is the factor:

$$\Phi_x \equiv \frac{Y_x - Y_{s,x}}{Y_c - Y_{s,c}} \approx \psi(\delta \cdot x) \quad (144)$$

Remember that (Eq. (114)):

$$Y_s(\alpha x) = \frac{\beta}{\alpha^2}\left[\frac{Bi \cdot \psi_s(\alpha x)}{\alpha \psi'_{s,\alpha} + Bi \cdot \psi_{s,\alpha}} - 1\right]$$

with:

$$\psi_s(\alpha x) = \frac{\sin[\alpha(x-x_0)]}{\alpha \cdot x} + \frac{x_0 \cdot \cos[\alpha(x-x_0)]}{x}$$

If the two temperature histories are simultaneously measured, Eqs. (141) and (144) allows us to the simultaneous determination of $\delta^2$ and $\alpha^2$. This is the ratio method extended to the simultaneous determination of the Biot number and the $\alpha^2$ number [47].

Consequently, if the thermal conductivity is known, from Eqs. (141) and (144), the surface heat transfer coefficient $h$ and the slope of the heat source $A_1$ can be indirectly determined.

## Other Indirect Determinations

General condition: It will be assumed from now on that the experiment is long enough, consequently the final temperature can be considered as the steady state temperature.

### Heat Transfer Coefficient

Condition: All the thermo-physical parameters and the two heat generation constants are known.

1) Transform the experimental time-temperature history at the thermal centre (pulp-stone contact) into a Table $\{Fo, \theta_i\}$, where $\theta_i = (T_i - T_s)/(T_0 - T_{ex})$
2) Obtain the dimensionless number of the heat respiration $\alpha^2 = q_1 R^2/k$
3) Obtain the linear regression to the linear portion of its graph in semi-logarithmic coordinates: $\ln \theta_c = M - N \cdot Fo \approx \ln[A] - N \cdot Fo$
4) Obtain the value $\delta^2$: $\delta^2 = N + \alpha^2$
5) And solve the Biot number in Eq. (116):

$$Bi = 1 + \frac{\delta^2 x_0 - m}{1 + m x_0} \qquad (145)$$

where $m$ is:

$$m \equiv \frac{\delta}{\tan[\delta(1-x_0)]} \tag{146}$$

Finally, clearing in the definition of Biot number, the surface heat transfer coefficient can be calculated:

$$h = \frac{Bi \cdot k}{R} \tag{147}$$

**Heat Generation Constants**

Condition: All the thermo-physical parameters and the surface heat transfer coefficient are known.

1. As all the thermo-physical parameters and the surface heat transfer coefficient are known, the Biot number can be calculated ($Bi = hR/k$).
2. From the transcendental boundary Eq. (116) the first eigenvalue $\delta$ can be calculated.
3. $\psi(\delta_i)$ and $A_0$ can be calculated from Eqs. (113) and (118).
4. As the steady state temperature $T_s$ is known, the $\{\theta - Fo\}$ table can be deduced and plotted, and from its linear portion, the linear regression can be obtained: $\ln \theta_c = M - NFo$.

Identifying to Eq. (136), $A$ and $\alpha^2$ can be determined:

$$A = e^M \tag{148}$$

$$\alpha^2 = \delta^2 - N \tag{149}$$

5. Taking into account Eqs. (117) and (118), and solving $\beta$:

$$\beta = N\left(1 - \frac{e^M}{A_0}\right) \tag{150}$$

6. Finally, clearing $q_0$, $q_1$ and $B_0$ in Eqs. (109), (110) and (111):

$$q_1 = \frac{\alpha^2 \cdot k}{R^2} \tag{151}$$

$$B_0 = \frac{\beta \cdot k \cdot \Delta T_0}{R^2} \tag{152}$$

and:

$$q_0 = B_0 - q_1 T_{ex} \tag{153}$$

## Indirect Measurement of Thermal Diffusivity and Surface Heat Transfer Coefficient

Condition: The thermal conductivity and the two heat generation constants $q_0$ and $q_1$ are known.

Similarly to Cuesta and Alvarez [13], the thermo-physical parameters and the Biot number can be determined both from the slope and the experimental lag factor $A$.

1. Obtain the dimensionless respiratory constants $\alpha^2$ and $\beta$.

As thermal conductivity $k$ is known, as well as the respiratory constants $q_0$ y $q_1$, substitute into Eqs. (109), (110) and (111) to obtain:

$$\alpha^2 = \frac{q_1 R^2}{k} = \text{Dimensionless slope of the heat source}$$

$$\beta = \frac{B_0 R^2}{k \cdot \Delta T_0} = \text{Dimensionless number of the heat source}$$

2. Obtain the experimental dimensionless difference $Y_{s,exp}$ (from the final portion of the time temperature table) and obtain the Table $\{\theta - t\}$ at the pulp-stone contact.

3. Obtain the linear regression to the linear portion of its graph in semi-logarithmic coordinates:

$$\ln\theta \approx M - S \cdot t$$

4. Identifying to Eq. (136), the lag factor is directly determined (Eq. (148)):

$$A = e^M$$

and:

$$S = \frac{a(\delta^2 - \alpha^2)}{R^2} \qquad (154)$$

5. On the other hand, be remembered that (Eqs. (117) and (118)):

$$A = A_0 \left(1 - \frac{\beta}{\delta^2 - \alpha^2}\right) \qquad (155)$$

$$A_0 = \frac{2Bi\psi(\delta)}{[\psi(\delta)]^2(\delta^2 + Bi^2 - Bi) - \delta^2 x_0{}^3} \qquad (156)$$

and:

$$\psi(\delta) = \frac{\sin[\delta(1 - x_0)] + \delta x_0 \cos[\delta(1 - x_0)]}{\delta} \qquad (157)$$

6. Bearing in mind that $\alpha^2 \& \beta \ll \delta^2$, for the sole purpose of establishing a first approximate value for $\delta$, the heat source will be omitted and the $\delta$ value corresponding to $A = e^M$ (Eq. 148) will be calculated as indicated in Cuesta and Alvarez [13]. This value will be considered as the starting point for the difference $(\delta^2 - \alpha^2)$ in the algorithm for calculating the coefficient of surface heat transmission and thermal diffusivity.

7. By introducing this $(\delta^2 - \alpha^2)$ and $\beta$ into Eq. (155), $A_0$ is determined.

$$A_0 = \frac{J}{\left(1 - \frac{\beta}{\delta^2 - \alpha^2}\right)} = \frac{e^M}{\left(1 - \frac{\beta}{\delta^2 - \alpha^2}\right)} = A_{0,0} \qquad (158)$$

8. Take a trial value for $\delta$.

9. The corresponding Biot number can be deduced using Eqs. (145) and (146).

10. Introduce this Biot number and the trial $\delta$ into Eqs. (156) and (157) to obtain $A_0$.

11. If $A_0 = A_{0,0}$ (within the deviation considered), then $\delta$, $Bi$ (i.e., the surface heat transfer coefficient) and the thermal diffusivity $a$ (by substituting in Eq. (154) have been simultaneously determined.

If $A_0 \neq A_{0,0}$ (within the deviation considered) then, by increasing or decreasing $\delta$, $A_0$ can be recalculated until the desired accuracy is obtained. In practice, using a simple algorithm to accelerate the convergence, no more than a few cycles of calculation are needed to determine $\delta$, $Bi$, and $a$.

12. With this $\delta$ calculate again $(\delta^2 - \alpha^2)$ to back to 7. When this difference repeats (within the deviation considered) the process is over and the surface heat transfer coefficient $h$ and the thermal diffusivity $a$ have been determined.

Consequently, if the thermal conductivity is known, from Eqs. (141) and (144), the surface heat transfer coefficient $h$ can be indirectly determined.

**Example V**

As just seen in the last algorithm, one of the applications of the first approximations is the indirect determination of the thermal diffusivity and the surface heat transfer coefficient. The condition is that the thermal conductivity and the two heat generation constants $q_0$ and $q_1$ are known.

A plum (*Prunus domestica*) weighing 125.77 g was stabilized (for 24 h) at $T_0 = 21.8\ °C$ and a 4-l bath was also stabilized into a storage chamber (for 24 h as well) whose set point temperature was 4.5 °C. At the time of introducing the plum into the bath, the water was stabilized at 4.6 °C, approximately.

The fruit approached a prolate spheroid with the larger diameter = 59.88 mm and the smaller diameter = 55.67 mm. After the complete experience, the stone was measured. Its dimensions were as follows: weight 2.060 g, larger diameter = 27.68 mm; intermediate diameter = 17.56 mm and smaller diameter = 11.11 mm.

The pulp weight is then $M = 0.12577 - 0.00206 = 0.12371$ kg.
The external measured volume is:

$$V = \frac{4}{3}\pi R_1 R_2 R_3 = 1.0452 \times 10^{-4} \text{ m}^3$$

Consequently, the equivalent radius is:

$$R = \sqrt[3]{R_1 R_2 R_3} = 0.02922 \text{ m}$$

The stone's volume is $2.8275 \times 10^{-6}$ m$^3$
The equivalent radius of the stone, $r_0 = 0.00877$ m
The measured pulp density was, therefore:

$$\rho = \frac{M_{tot} - M_{stone}}{V_{tot} - V_{stone}} = 1216.53 \text{ kg m}^{-3}$$

The ratio stone/fruit radius is:

$$x_0 = \frac{r_0}{R} = 0.30$$

From ASHRAE [25], (p. 19.12 Table 5 and p. 19.18 Table 7), density, thermal conductivity and thermal diffusivity are:

$$\rho = 1219 \text{ kg m}^{-3}$$

$$k = 0.375 \text{ W} \cdot \text{m}^{-1} \cdot \text{K}^{-1}$$

$$a = 1.2 \text{ m}^2 \cdot \text{s}^{-1}$$

And from ASHRAE [12], the linear regressions of the heat of respiration constants (average values) between 0 and 20 °C are:

$$q_0 = 5.030 \times 10^{-3} \text{ J} \cdot \text{kg}^{-1} \cdot \text{s}^{-1}$$

$$q_1 = 2.658 \times 10^{-3} \text{ J} \cdot \text{kg}^{-1} \cdot \text{s}^{-1} \cdot \text{K}^{-1}$$

As the units for $q_0$ and $q_1$ in this chapter are $\text{J} \cdot \text{m}^{-3} \cdot \text{s}^{-1}$ and $\text{J} \cdot \text{m}^{-3} \cdot \text{s}^{-1} \cdot \text{K}^{-1}$ respectivelly, by multiplying the above $q_0$ and $q_1$ by the density of the pulp:

$$q_0 = 6.119 \text{ J} \cdot \text{m}^{-3} \cdot \text{s}^{-1}$$

$$q_1 = 3.234 \text{ J} \cdot \text{m}^{-3} \cdot \text{s}^{-1} \cdot \text{K}^{-1}$$

That is:

$$B_0 = q_0 + q_1 T_{ex} = 20.890 \text{ J} \cdot \text{m}^{-3} \cdot \text{s}^{-1}$$

$$\alpha^2 = \frac{q_1 R^2}{k} = 7.363 \times 10^{-3}$$

$$\beta = \frac{B_0 R^2}{k \cdot \Delta T_0} = 2.760 \times 10^{-3}$$

From the original time (s) – temperature (°C) tables, the average temperature of the water was $T_{ex} = 4.568$ °C, and the average final temperature of the plum at the stone/fruit interface was approximately 4.594 °C That is, $Y_s = (T_s - T_{ex})/\Delta T_0 \approx 1.515 \times 10^{-3}$.

Otherwise, from the original time (s) – temperature (°C) tables, it is possible to derive the $\{t - \theta\}$ table, which is shown in Figure 14a on a semi-logarithmic scale, where the linear portion is clearly visible. The linear regression is:

$$\ln \theta_c = M - S \cdot t \approx 0.3243 - 1.0255 \times 10^{-3} \cdot t$$

With a coefficient $R^2 = 0.999$.
That is:

$$A = e^M = 1.3830$$

The slope $\delta^2$ corresponding to this value $A$ regardless of the heat source is:

$$\delta^2 = 5.0304$$

Which will be considered as a starting point for the difference $(\delta^2 - \alpha^2)$.

Consequently, substituting into Eq. (158) and clearing $A_0$:

$$A_0 = \frac{1.3830}{\left(1 - \frac{2.760 \times 10^{-3}}{5.0304}\right)} = 1.3838 = A_{0,0}$$

Then, following steps 8 to 11 of the process described previously, the following values are obtained:

$$\delta = 2.245$$

Corresponding to the Biot Number:

$$Bi = 2.514$$

Therefore, clearing $a$ in Equation (154) we get to:

$$a = 1.740 \times 10^{-7} \text{ m}^2\text{s}^{-1}$$

And, clearing $h$ in Biot number:

$$h = 32.26 \text{ Jm}^{-2}\text{s}^{-1}\text{K}^{-1}$$

As the Biot number is already known, the roots of the transcendental boundary equation (Eq. 116) and the constants of the serial expansion (Eqs. 117 and 118) can be calculated. Consequently, the complete theoretical time-temperature curve of this experiment can be drawn and compared with the experimental values. The standard error between experimental and theoretical $Y$ curves is $\pm 0.27$ °C.

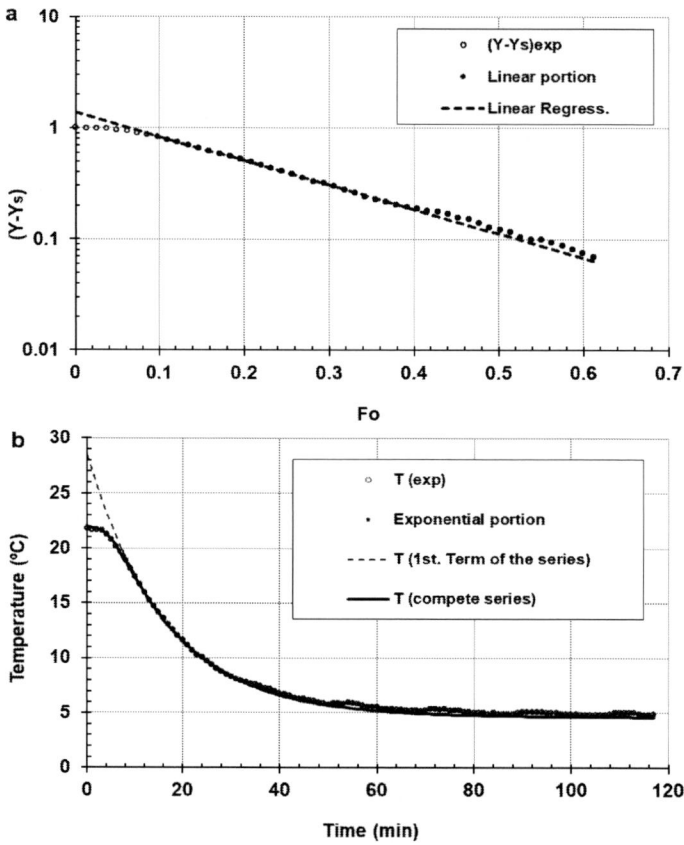

Figure 14. (a) Linear portion and linear regression (in semi logarithmic scale); (b) Experimental kinetic of temperature, exponential portion, first term of the series and complete series.

The result of the complete series corresponding to this Biot number is shown in Figure 14b (unbroken line), along with the experimental values (empty circles).

# REFERENCES

[1] Jakob, M. (1949). *Heat Transfer*. John Wiley & Sons, Inc., Hoboken, New Jersey, Chapter 10, Section 10-4.

[2] Carslaw, H. S., & Jaeger, J. C. (1959). *Conduction of Heat in Solids*, 2nd. ed. Clarendon Press, Oxford, pp. 246–250.

[3] Gröber, H., Erk S., & Grigull, U. (1961). *Fundamentals of Heat Transfer*, 3rd ed. McGraw-Hill, New York.

[4] Eckert, E. R., & Drake, R. M. (1974). *Heat and Mass Transfer*, McGraw-Hill, New York.

[5] Heldman, D. R., & Lund, D. B. (1992). *Handbook of Food Engineering*. Marcel Dekker, New York.

[6] Pflug, I. J., Blaisdell, J. L., & Kopelman, J. (1965). Developing temperature-time for objects that can be approximated by sphere, infinite plate or infinite cylinder. *ASHRAE Transactions, 71*, 238-248.

[7] Smith, R. E., Nelson, G. L., & Henrickson, R. L. (1967). Analyses on transient heat transfer from anomalous shapes. *Transactions of the ASAE, 10*(2), 236-245.

[8] Clary, B. L., Nelson, G. L., & Smith, R. E. (1968). Heat transfer from hams during freezing by low-temperature air. *Transactions of the ASAE, 11*(4), 496-499.

[9] Cleland, A. C., & Earle, R. L. (1982). A simple method for prediction of heating and cooling rates in solids of various shapes. *International Journal of Refrigeration, 5*, 98-106.

[10] Cuesta, F. J., Lamtia, M., & Moreno, J. (1990). Graphical calculation of half-cooling times. *International Journal of Refrigeration, 13*, 317-324.

[11] Cuesta, F. J., & Lamúa, M. (1995). Asymptotic modelling of the transient regime in heat conduction in solids with general geometry. *Journal of Food Engineering, 24*, 295–320.

[12] ASHRAE (2010). *ASHRAE Handbook – Refrigeration (SI Edition)*. American Society of Heating, Refrigerating and Air Conditioning Engineers, Inc., 1791 Tullie Circle, N. E., Atlanta, GA.

[13] Cuesta, F. J., & Alvarez, M. D. (2017). Mathematical modeling for heat conduction in stone fruits. *International Journal of Refrigeration, 80*, 120-129.

[14] Cuesta, F. J., & Alvarez, M. D. (2019). Theoretical Fourier series solution for heat conduction in stone fruits with internal heat source linearly reliant on temperature. *In Agricultural Research Updates*, Volume 27, P. Gorawala & S. Mandhatri (Eds.), Nova Science Publishers, Inc, New York.

[15] Awuah, G. B., Ramaswamy, H. S., & Simpson, B. K. (1995). Comparison of two methods for evaluating fluid-to-surface heat transfer coefficients. *Food Research International, 28*(3), 261–271.

[16] Erdoğdu, F. (2008). A review on simultaneous determination of thermal diffusivity and heat transfer coefficient. *Journal of Food Engineering, 86*, 453–459.

[17] Kondjoyan, A. (2006). A review on surface heat and mass transfer coefficients during air chilling and storage of food products. *International Journal of Refrigeration, 29*, 863–875.

[18] Erdoğdu, F., Linke, M., Praeger, U., Geyer, M., & Schlüter, O. (2014). Experimental determination of thermal conductivity and thermal diffusivity of whole green (unripe) and yellow (ripe) Cavendish bananas under cooling conditions. *Jounal of Food Engineering, 128*, 46–52.

[19] Erdoğdu, F. (2005). Mathematical approaches for use of analytical solutions in experimental determination of heat and mass transfer parameters. *Journal of Food Engineering, 68*, 233–238.

[20] Cinquanta, L., Di Matteo, M., & Estia, M. (2002). Physical pretreatment of plums (*Prunus domestica*). Part 2. Effect on the quality characteristics of different prune cultivars. *Food Chemistry, 79*(2), 233–238.

[21] Hernández, F., Pinochet, J., Moreno, M. A., Martínez, J. J., & Legua, P. (2010). Performance of Prunus rootstocks for apricot in Mediterranean conditions. *Scientia Horticulturae, 124*, 354–359.

[22] Jiménez-Jiménez, F., Castro-García, S., Blanco-Roldán, G. L., Ferguson, L., Rosa, U. A., & Gil-Ribes, J. A. (2013). Table olive

cultivar susceptibility to impact bruising. *Postharvest Biology and Technology, 86*, 100–106.

[23] Cuesta, F. J., Lamúa, M., & Alique, R. (2012). A new exact numerical series for the determination of the Biot number: application for the inverse estimation of the surface heat transfer coefficient in food processing. *International Journal of Heat and Mass Transfer, 55*, 4053–4062.

[24] Uyar, R., & Erdoğdu, F. (2012). Numerical evaluation of spherical geometry approximation for heating and cooling of irregular shaped food products. *Journal of Food Science, 77*, E166–E175.

[25] ASHRAE (1998). Thermal Properties of Foods. *ASHRAE Handbook – Refrigeration (SI)*. American Society of Heating, Refrigerating and Air Conditioning Engineers, p. 8.2. (Chapter 8).

[26] Carson, J. K. (2006). Review of effective thermal conductivity models for foods. *International Journal of Refrigeration, 29*, 958–967.

[27] Carson, J. K., Lovatt, S. J., Tanner, D. J., & Cleland, A. C. (2005). Thermal conductivity bounds for isotropic, porous materials. *International Journal of Heat and Mass Transfer, 48*, 2150–2158.

[28] Carson, J. K., Lovatt, S. J., Tanner, D. J., & Cleland, A. C. (2006). Predicting the effective thermal conductivity of unfrozen, porous foods. *Journal of Food Engineering, 75*, 297–307.

[29] Choi, Y., & Okos, M. R. (1986). Effects of temperature and composition on the thermal properties of foods. In: Maguer, L., Jelen, P. (Eds.), *Food Engineering and Process Applications: Transport Phenomenon*, vol. 1. Elsevier, New York, pp. 93–116.

[30] Marcotte, M., Taherian, A. R., & Karimi, Y. (2008). Thermophysical properties of processed meat and poultry products. *Journal of Food Engineering, 88*, 315–322.

[31] Maroulis, Z. B., Krokida, M. K., & Rahman, M. S. (2002). A structural generic model to predict the effective thermal conductivity of fruits and vegetables during drying. *Journal of Food Engineering, 52*, 47–52.

[32] Murakami, E. G., & Okos, M. R. (1988). Measurement and prediction of thermal properties of foods. In: Singh, R. P., Medina, A. G. (Eds.), *Food Properties and Computer-aided Engineering of Food Processing Systems (Part 1)*. NATO ASI series. Kluwer Academic Publishers, Boston, pp. 3–48.

[33] Guillén, R., Heredia, A., Felizón, B., Jiménez, A., Montaño, A., & Fernández-Bolaños, J. (1992). Fibre fraction carbohydrates in Olea europaea (Gordal and Manzanilla var. *Food Chemistry. 44*, 173–178.

[34] Ongen, G., Sargın, S., Tetik, D., & Köse, T. (2005). Hot air drying of green table olives. *Food Technology and Biotechnology, 43*, 181–187.

[35] Wang, J. F., Carson, J. K., North, M. F., & Cleland, D. J. (2006). A new approach to the modeling of the effective thermal conductivity of heterogeneous materials. *International Journal of Heat and Mass Transfer, 49*, 3075–3083.

[36] Campañone, L. A., Giner, S. A, & Mascheroni, R. H. (2002). Generalized model for the simulation of food refrigeration. *International Journal of Refrigeration, 25*, 975–984.

[37] Tanner, D. J., Cleland, A. C., Opara, L. U., Robertson, T. R. (2002a). A generalized mathematical modelling methodology for the design of horticultural food packages exposed to refrigerated conditions. Part 1. Formulation. *International Journal of Refrigeration, 25*, 43–53.

[38] Tanner, D. J., Cleland, A. C., & Opara, L. U. (2002b). A generalized mathematical modelling methodology for the design of horticultural food packages exposed to refrigerated conditions. Part 2. Heat transfer modelling and testing. *International Journal of Refrigeration, 25*, 43–53.

[39] Dincer, I. (1994). Unsteady heat-transfer analysis of spherical fruit to air flow. *Energy, 19*, 117–123.

[40] Dincer, I. (1997). *Heat Transfer in Food Cooling Applications* (Taylor and Francis, Washington, Chapter 5), pp. 204-233.

[41] Meffert, H. F. Th., Rudolphij, J. W., & Rooda, J. W. (1971). *Proceedings of the XIII Congress of Refrigeration, 2*, 379-385.

[42] Mendonça, S. L. R., Filho, C. R. B., & da Silva, Z. E. (2005). Transient conduction in spherical fruits: method to estimate the thermal conductivity and volumetric thermal capacity. *Journal of Food Engineering*, *67*, 261–266.

[43] Wang, S., Tang, J., Cavalieri, R. P. (2001). Modeling fruit internal heating rates for hot air and hot water treatments. *Postharvest Biology* and *Technology*, *22*, 257–270.

[44] Sadashive Gowda, B., Narasimham, G. S. V. L., & Krishna Murthy, M. V. (1997). Forced-air precooling of spherical foods in bulk: a parametric study. *International Journal of Heat and Fluid Flow*, *18*, 613–624.

[45] Kole, N. K., & Prasad, S. (1994). Respiration rate and heat of respiration of some fruits under controlled atmosphere conditions. *International Journal of Refrigeration*, *17*, 199–204.

[46] Rao, N., Flores Rolando, A., & Gast Karen, L. B. (1993). Mathematical relationships for the heat of respiration as a function of produce temperature. *Postharvest Biology and Technology*, *3*, 173–180.

[47] Xu, Y., & Burfoot, D. (1999). Simulating the bulk storage of foodstuffs. *Journal of Food Engineering*, *39*, 23-29.

[48] Cuesta, F. J., & Lamúa, M. (2009). Fourier series solution to the heat conduction equation with an internal heat source linearly dependent on temperature: Application to chilling fruit and vegetables. *Journal of Food Engineering*, *90*, 291-299.

[49] Rey Pastor, J., & De Castro-Brzezicki, A. (1958). *Funciones de Bessel*, Dossat, Madrid, pp. 231. [*Bessel functions*]

In: Understanding Heat Conduction
Editor: William Kelley
ISBN: 978-1-53619-182-0
© 2021 Nova Science Publishers, Inc.

*Chapter 2*

# SENSITIVITY OF NUMERICAL MODELING TECHNIQUE FOR CONJUGATE HEAT TRANSFER INVOLVING HIGH SPEED COMPRESSIBLE FLOW OVER A CYLINDER

*Laurie A. Florio*[*]
Armaments Technology and Evaluation Division,
US ARMY Armament Graduate School,
Picatinny Arsenal, NJ, US

## ABSTRACT

The surface heating due to conjugate heat transfer is often the main source of conduction heat transfer through solid materials moving through or impinged upon by high speed compressible flows. High temperature gradients and large heat fluxes at the fluid-solid interfaces develop under such conditions. Since high temperatures and high temperature gradients affect the solid material properties and thus material strength, an understanding of these conjugate heat transfer

---

[*] Corresponding Author's E-mail: laurie.a.florio.civ@mail.mil.

phenomena is important. The compressible flow around a fixed cylinder or due to a moving cylinder is a well-established baseline for comparison. Numerical models can be used to investigate and predict the resulting thermal conditions for such flow and heat conduction effects, but the numerical modeling methods must be attuned to properly capture the conditions unique to this type of flow. Among the factors influencing the numerical results are the mesh size, the time step and time discretization, various discretization methods including gradient calculations and gradient limiters as well as the turbulence model and associated parameters and options. When a moving object requiring some mesh motion is added to the simulation, additional options become available. A sliding type of mesh motion treats a particular zone as a rigid body, moving the cylinder along a linear path, for quadrilateral/hexahedral element types. Remeshing algorithms deform elements and then split or coalesce elements as an object moves through the mesh, generally requiring a triangular/tetrahedral element type that may be more numerically diffuse. An overset mesh, where the mesh associated with the cylinder moves over a fixed background mesh, can be used to model the movement of the cylinder in a simulation with quadrilateral/hexahedral elements, without remeshing. However, the accuracy of the overset method for high temperature gradient conditions is not well defined. The sensitivity of conditions that develop as a result of high speed compressible flow over a cylinder to changes in the mesh, numerical models, and mesh motion methods is explored in this work with comparisons made to published data. All studies are conducted for a 0.0762m diameter steel cylinder in air with the pressure, temperature, and heat flux around the outer surface of the cylinder compared for the various models. A Mach 6.46 flow is studied first with comparisons made to published data. Then, the sensitivity of the flow conditions to the modeling methods or model related parameters is investigated for the same system. In the final part of the study, three different methods of applying mesh motion are implemented for subsonic, near sonic, and supersonic flow conditions to examine the effect of the mesh motion method on the predicted flow conditions and the relationship to the flow speed. These studies provide the information needed to better select the modeling methods to use for conjugate heat transfer analysis with high speed compressible flow under a given set of conditions.

**Keywords**: conjugate heat transfer, computational fluid dynamics, cylinder, high speed compressible flow

# INTRODUCTION

As high speed compressible flow impinges upon a solid body in its path, kinetic energy is converted to pressure and thermal energy as the fluid flow slows. The resulting flow and thermal conditions at the fluid-solid interface determine the forces acting on the solid body, the heat transfer into the solid body, and the temperatures that develop within the solid. The temperature levels and pressure values reached are typically highest near the stagnation zones, where fluid motion has been decelerated to the largest extent. Particularly for the high speed compressible flow, the gas velocity, pressure, and temperature gradients are high at the fluid-solid interface surfaces. The ability to estimate the flow induced loading and conjugate heat transfer is critical to ensuring the structural components can withstand the mechanical and thermal loading that result. Computational models offer a controlled means of exploring the flow and temperature field conditions caused by the high speed fluid flow and the conduction into the solid objects upon which the flow impinges. However, a number of options and approaches are available to model these high speed conjugate heat transfer-flow phenomena. The numerical methods must be attuned to capturing the flow phenomena typical for these high speed compressible flow-conjugate heat transfer conditions. The solver type and discretization methods, along with the mesh type, mesh size, timestep, turbulence model, and turbulence modeling options including wall treatment and compressibility effects can affect the flow and temperature field predictions. This work investigates the sensitivity of the interface pressure, temperature, and heat flow conditions, specifically, to the variation in the modeling method options and parameters for the high speed air flow over a cylinder.

Flow over a cylinder is a well-established baseline that is used in this work for investigating the sensitivity of the results of simulations of the conjugate heat transfer that occurs as a solid body is exposed to high speed compressible flow to changes in the modeling methods and parameters. One of the most fundamental works for high speed compressible flow over a cylinder is that by Wieting (Wieting 1987). Wieting performed experiments on shock wave impingement on a fixed 0.0762 m diameter cylinder. The aerodynamic and aero-thermal loading is captured over a range of flow speeds and flow angles varied relative to a fixed solid steel cylinder, including a Mach 6.46 flow at an angle normal to the cylinder axis. The distribution of the local heat rate and pressure field at the cylinder surface is tabulated as these flow parameters are altered. The highest pressures and heat fluxes occur near the stagnation point, and the pressure, heat flux, and temperature values decay moving around the surface of the cylinder. In the rear wake region, the flow velocities are lower and the cylinder surfaces are more protected from the high gradients in the flow and thermal conditions near the stagnation point and near the shock that forms on the upstream side of the cylinder. Hence, the heat conduction through the solid material that results from the heat load applied at the fluid-solid interface varies significantly around the circumference of the cylinder. Application of a constant temperature boundary condition at the cylinder's outer surface may not properly estimate the heat load experienced by the cylinder. In addition to allowing for the development of radial and angular variation of the temperature within the cylinder, the models should account for temperature dependent properties in the solid as well as in the fluid due to the wide range of temperature values that might be present. Wieting (Wieting 1987) notes the models that incorporate temperature varying properties in the solids more closely followed experimental data obtained in the study. The results of the studies of Wieting also are the reference to which a number of later studies compare.

Among the studies working to computationally replicate the flow and conjugate heat transfer for the hypersonic flow over a fixed cylinder studies of Wieting are the works of Zhao et al. (Zhao, et al. 2011) and Murty et al. (Murty, Manna and Chakraborty 2012). Zhao et al. performed thermal-fluid coupled computational fluid dynamics analyses with a hollow cylinder with a fixed internal cylinder surface temperature, with the inner surface situated a sufficient radial distance from the outer surface so that the heat transfer does not reach the inner surface of the cylinder over the time period studied and a constant inner cylinder surface temperature can be supplied. In this study, the flow field is assumed to reach equilibrium without any thermal conduction into the cylinder solid material. When the flow is established, then the heat flow into the cylinder is allowed, but a clear timeline of when this flow is established is not provided. The thermal properties of the cylinder steel material are not explicitly stated. The predicted pressures and heat flux, normalized by their respective stagnation values, as calculated by Zhao et al. (Zhao, et al. 2011) compare favorably with the heat results of Wieting (Wieting 1987).

The study by Murty (Murty, Manna and Chakraborty 2012) also involves a computational model of the flow over a fixed cylinder under hypersonic conditions, again following the experiments conducted by Wieting. Mutry provides more details on the numerical methods selected for the study. A density based implicit coupled solver is used with the Advection Upstream Splitting Method selected and second order central differencing for the discretization method. Menter's SST turbulence model is applied for the cylindrical conjugate heat transfer problem. While mention is made of gas species, no information on the gas species conditions or the specific gas species involved are explicitly provided. In this study, no specific material properties for the steel or the initial conditions in the steel cylinder are given. The differences that arise with the constant vs. variable solid properties are reported. The shape of the shock patterns as well and the normalized surface pressure and heat flux show good comparison to the experimental data of Wieting. Murty (Murty, Manna and Chakraborty 2012) notes that the deformation of the cylinder is small and the structural deformation effects can be neglected for the conditions studied. Murty (Murty, Manna and Chakraborty 2012) does suggest, however, that two dimensional heat conduction becomes more important at material interfaces for the aero-thermal problems, perhaps indicating considering solely radial heat conduction may not be sufficient for the current system, particularly with the non-uniform thermal loading.

In both of these studies, the effects of the selection of the computational fluid dynamics modeling methods, settings, or parameters on the temperatures, pressures, and heat flux that develop at the fluid-solid interface for high speed compressible flow-solid body interactions have not been addressed. In this current work, the influence of a number of computational fluid dynamics modeling methods and related parameters on the overall flow and temperature field features as well as the distribution of the temperature, pressure, and heat flux around a solid cylinder held in high speed flow are explored. After comparing the current baseline model to a bench mark Wieting case, a series of studies are conducted, modifying the basic flow computational models selected and modeling parameters to obtain an understanding of how sensitive the model predicted flow and thermal loading on the cylinder is to the selection of the modeling method. This first part of the investigation provides information on the sensitivity of the numerical results to the modeling methods and settings. Then, with the model settings fixed, the flow and thermal conditions that develop with a moving cylinder for three different cylinder velocities and three different means of computationally moving the cylinder through the gas, are investigated and the results compared. The second portion of the study can be used to indicate the moving mesh methods that may be more viable to simulate the motion of the cylinder through the gas. The results of these studies are reported in this work. This work aims to improve the overall understanding of the conjugate heat transfer that takes place within the cylinder exposed to high speed gas flow and the effects of the selection of different methods for computationally modeling these events.

## METHODS

Finite volume, computational fluid dynamics based methods are used to determine the velocities, temperatures, pressures, and turbulence parameters in the compressible fluid and the resulting temperatures in the solid cylinder. Further information on the system investigated and the specific modeling methods and parameters used are described below.

## System Investigated

The baseline system investigated is a fixed cylinder with the high speed flow moving over the cylinder as depicted in Figure 1. The cylinder has a diameter of 0.0762 m. Replicating the work of Zhao (Zhao, et al. 2011) and Murty (Murty, Manna and Chakraborty 2012), the thickness of the cylinder is 0.0127 m, assuming over the time period studied, the heat flow has not yet reached this position. The free stream gas temperature is 241.5K, the freestream pressure is 648.1Pa, and the Mach number is 6.46. The validation data is taken at 5s.

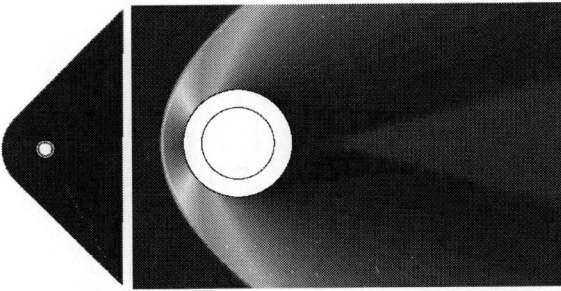

Figure 1. Conjugate heat transfer system considered showing solid cylinder.

With the basic modeling method comparing favorably with published data, the same baseline system geometry is utilized and the same freestream conditions are applied, but now the modeling methods are modified. Further information will be described in the modeling methods section later in this work.

Finally, instead of a freestream flow specified, the cylinder is moved through the fluid at a given velocity and the subsequent flow and temperature fields that develop are analyzed. The initial pressure of 101325Pa and initial temperature of 300K are assigned to the gas and an initial temperature of 300K is assigned to the cylinder material. In these cases, the cylinder velocity is accelerated from rest to a constant velocity in 1e-05 seconds. Constant velocities of 250 m/s, 500 m/s, and 1000 m/s are set for the cylinder motion. In this way, a subsonic, near sonic, and supersonic flow regime are investigated for the same geometric configuration. For each of these velocities, three methods of cylinder motion are implemented:

1. A sliding mesh is used as in Figure 2a, where the mesh zone containing the cylinder slides between the upper and lower fluid zones, carrying the cylinder downstream. This method of mesh motion requires the use of quadrilateral or hexahedral element types as elements are added in a cell layer at a specified surface and elements are destroyed at another surface, maintaining the same total volume and requiring a rigid body motion of the mesh zone.

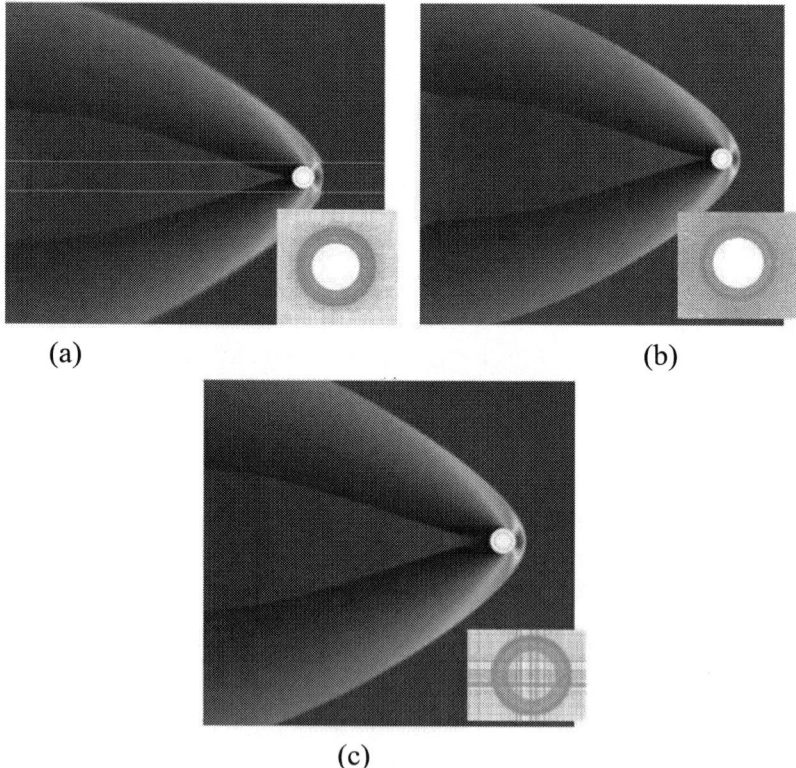

Figure 2. Mesh set-up for the three mesh motion methods; (a) Sliding; (b) Remeshing; (c) Overset.

2. The second method is a remeshing method (see Figure 2b). The cylinder moves through the fluid deforming the nearby mesh, stretching or compressing the mesh. When the element size reaches above a critical value, large elements are divided into new elements and when the element size reaches below a critical value, the smaller elements are added together to create an element of larger size. The remeshing process helps to limit severe mesh deformation or collapse and allows for large scale object motion in six degrees of freedom.
3. The last moving mesh method tested is the overset method. Here a background mesh is developed and this background mesh remains fixed. The moving cylinder mesh sits over this background mesh as in Figure 2c. This overset mesh moves over the background mesh with interpolation used to transmit the flow conditions to the background mesh. Three dimensional, six-degree-of-freedom motion can be assigned for this system as well. The type of mesh is not restricted, but interpolation may reduce the accuracy of high gradient conditions.

## Governing Equations

The following equations govern the conditions in the fluid and in the solid.

The specific governing equations utilized in this work are the conservation of mass, momentum, and energy in the fluid.

The continuity equation is: (White 1998):

$$\frac{\partial \rho}{\partial t} + \nabla \cdot (\rho \boldsymbol{V}) = 0 \qquad \text{Eq.(1)}$$

The conservation of momentum is (White 1998):

$$\frac{\partial(\rho \mathbf{V})}{\partial t} + \nabla \cdot (\rho \mathbf{VV}) = \nabla \cdot (\boldsymbol{\tau}) - \nabla p$$

$$\tau = \mu \left[ \nabla \mathbf{V} + \nabla \mathbf{V}^T - \frac{2}{3} \nabla \cdot \mathbf{V} \mathbf{I} \right] \qquad \text{Eq. (2)}$$

In the equation above, $\tau$ is the stress tensor. (White 1998) The conservation of energy is:

$$\rho Dh/Dt = Dp/Dt + \nabla \cdot (k\nabla T) + \tau_{ij} \partial v_i/\partial x_j \qquad \text{Eq. (3)}$$

In the solid cylinder, only heat conduction is considered where the properties are those of the solid and so the conservation of energy is:

$$\rho c_p \frac{\partial T}{\partial t} = \nabla \cdot (k \nabla T) \qquad \text{Eq. (4)}$$

## Material Properties

The gas utilized in the flow over the cylinder is air and stainless steel is the material for the solid cylinder. The air is treated as an ideal gas with a polynomial temperature varying specific heat and a kinetic theory based viscosity and thermal conductivity.

For the air, the molecular weight is 28.966 with an ideal gas equation of state implemented. The Leonard-Jones Characteristic Length is 3.711 and the Leonard-Jones Energy Parameter is 78.6. The temperature varying specific heat in J/kg-K is:

For T[100K,1000K]; Cp(T)=1161.482-3.368819T+
0.01485511$T^2$-5.034909e-05$T^3$+9.92857e-08$T^4$-1.111097e-10$T^5$+6.540196e-14$T^6$-1.573588e-17$T^7$      Eq. (5)

For T(1000K,3000K]; Cp(T)=-7069.814+33.70605T-
0.0581275$T^2$+5.421615e-05$T^3$-2.936679e-08$T^4$+9.237533e-12$T^5$-1.565553e-15$T^6$+1.112335e-19$T^7$      Eq. (6)

For the steel, the baseline properties used are a density of 7800kg/m³, a specific heat of 586 J/kg-K at 300K, 670 J/kg-K at 500K, and 1050 J/kg-K at 1000K, and a thermal conductivity of 15.7 W/m²K at 300K, 20.8 W/m²K at 500K, and 29.4 W/m²K at 900K with linear interpolation at temperatures between these values.

## Modeling Method Studies

All simulations in this investigation involve a two dimensional, transient finite volume based computational fluid dynamics based models. The commercial software FLUENT© is used to carry out the numerical simulations. The baseline modeling method is outlined here. A pressure based coupled solver is applied. Second order time discretization with Second Order Upwinding, Least Squares Cell Based Gradients, a k-ω, SST turbulence model with the Compressibility Effects off, and the solver TVD (Total Variation Diminishing) flux limiter off. For these simulations, the model is run until a time of 5 seconds is reached with a timestep 5.0e-05s. The simulations were run on 72 processors for approximately 60 hours to reach the 5 second time.

The same fixed cylinder system is investigated when varying the modeling methods and parameters, but a shorter time period of 0.4s is utilized. In these studies of the effect of modeling methods or parameters are investigated. The items altered in the modeling technique are included in Table 1. For these studies a time step of 1e-07s was used as the baseline with the simulations run on 72 processors for approximately 18 hours.

For the final portion of the simulations, where the effects of the mesh motion method and cylinder velocity are investigated for a moving cylinder, the baseline modeling methods are applied, but with the transient discretization switched to first order to accommodate all of the mesh motion techniques. These simulations, particularly for the remeshing method, required longer run-times with a timestep of 1.0e-06s utilized for these studies. All simulations ended when the cylinder had moved by approximately 0.4m.

**Table 1. Cases for sensitivity to modifications to the modeling method**

| Item | Description |
|---|---|
| A.Time discretization method | Baseline = Second Order, Modified = First Order |
| B.Time step | Baseline = 1.0e-07s, Modified = 2.0e-07s |
| C.Upwinding | Baseline = Second Order Upwinding; Modified = First Order Upwinding |
| D.Gradient Calculations | Baseline = Least Sqaures Cell Based; Modified = Green Gauss Node Based |
| E.Flux Limiter | Baseline = Off; Modified = On |
| F.Turbulence - Compressibility | Baseline = $\kappa$ - $\omega$ SST, No Compressibility Effects; Modified = $\kappa$- $\omega$ SST Compressibility Effects |
| G.Turbulence – $\kappa$- $\omega$ variation | Baseline= $\kappa$ -$\omega$- SST; Modified = $\kappa$ - $\omega$ -standard |
| H.Turbulence – $\kappa$ -$\varepsilon$ variation | Baseline = $\kappa$ -$\omega$-SST; Modified = $\kappa$ -$\varepsilon$-non-equilibrium wall |
| I.Turbulence – $\kappa$ -$\varepsilon$ variation | Baseline = $\kappa$ -$\omega$-SST; Modified = $\kappa$ -$\varepsilon$-enhanced wall treatment |

## Model Validation

The results of the standard cylinder geometry case are compared to the results in Zhao (Zhao, et al. 2011) which reported on a numerical model replicating the conditions studied by Wieting. An initial baseline mesh with the mesh graded in both the solid and fluid near the fluid-solid interface was generated. In order to establish a refined mesh near the shock boundary, the initial model was run through basic flow feature development with the given base mesh and then the mesh was refined along pressure gradients. The refinement assists in better resolving the flow conditions near the shock. After running a model with this mesh, the base mesh was refined by increasing the number of elements around the circumference of the cylinder by up to 50% and also decreasing the minimum size of the first element from the surface and the subsequently graded elements at the fluid-solid interface by a factor of up to 2.5. A similar procedure is used to then refine the mesh at the shock interface based on the pressure gradient. The changes in the heat flux at the fluid solid interface due to the alteration in the mesh are tracked for the different mesh cases at the same 5s time. The stagnation location normalized pressure and heat flux distribution around the outer cylinder surface are compared to the results in Zhao (Zhao, et al. 2011). The final mesh showing the area of refinement at the shock boundary are provided in Figure 3.

Next, the local data for the normalized pressure and heat flux from the current model are compared to the published results in Zhao and Murty. Figure 4 shows the normalized heat flux data for three of the meshes examined. The heat flux required two refinements of the mesh before little change in the distribution is observed (under 2% of the maximum value). The accurate heat flux value requires a more refined mesh than for the pressure distribution near the interface. For the final refined mesh, the heat flux distribution around the cylinder predicted by the model and reported by Zhao are provided in Figure 5. Good agreement between current local data and the published local data is found.

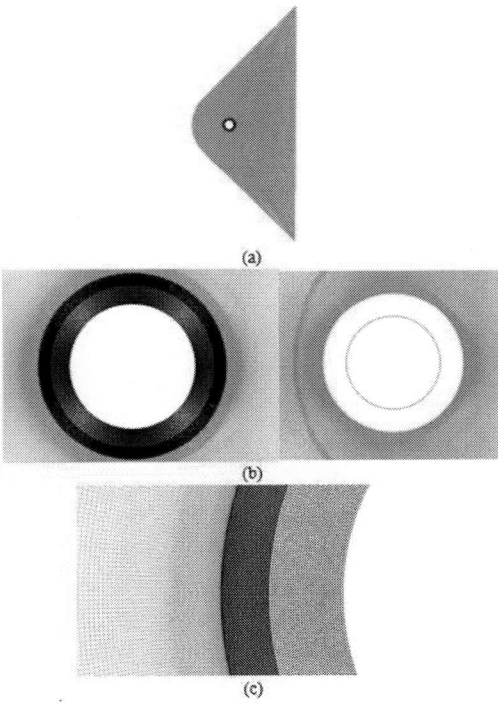

Figure 3. Mesh: (a) overall (b) cylinder (c) near cylinder.

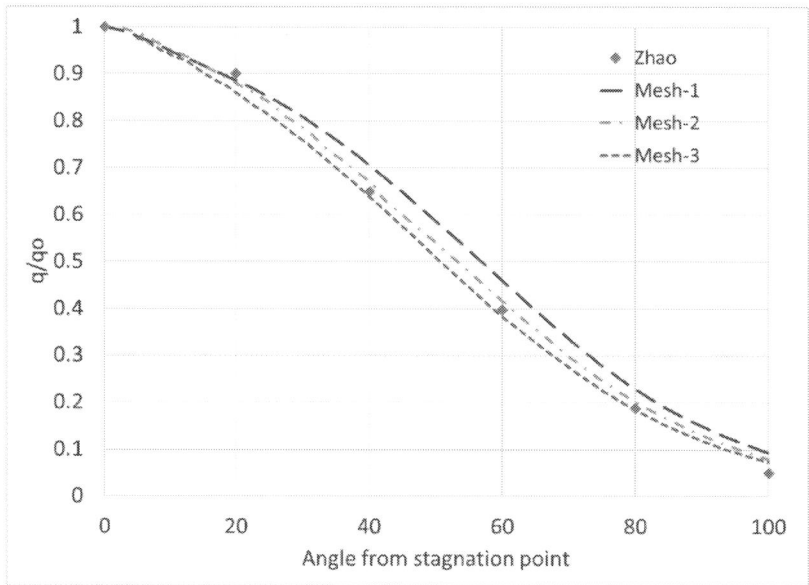

Figure 4. Stagnation normalized heat flux with mesh refinement, compared to Zhao data (Zhao, et al. 2011).

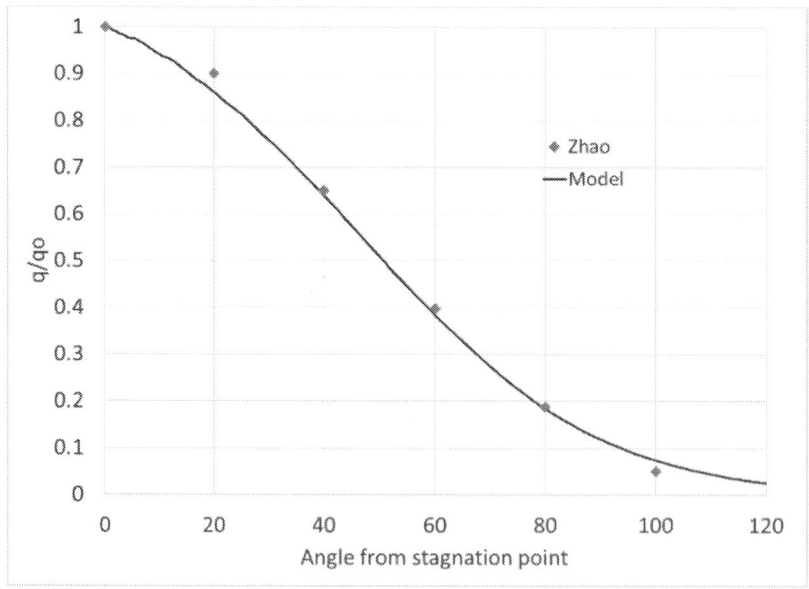

Figure 5. Comparison of the stagnation normalized heat flux predicted by the model and from Zhao (Zhao, et al. 2011).

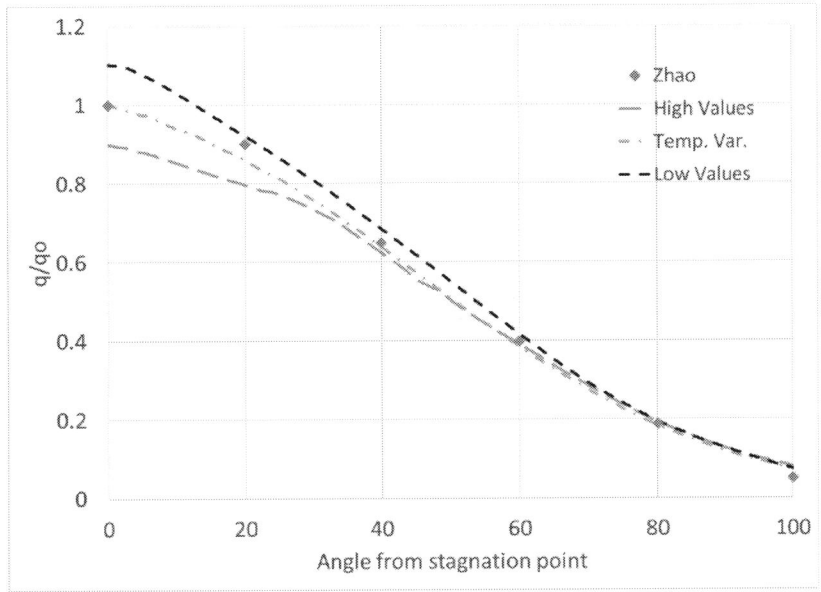

Figure 6. Cylinder heat flux variation with steel material properties.

Since the thermal properties of the stainless steel cylinder were not distinctly specified in the literature, the sensitivity of the thermal conditions at 5 seconds to varied thermal properties of the cylinder was investigated. Constant thermal properties for the steel are applied, setting a constant value equal to the lower limit (marked low values) and the upper limit (marked high values) of both the specific heat and thermal conductivity values from the description of the thermal properties of the steel earlier in this work. The results showed the stagnation heat flux can be altered by about 10% with the variation in the thermal properties of the cylinder material as seen in Figure 6, where the heat flux has been normalized by the peak value for the temperature varying case. Hence, the proper material properties are important in modeling the conductive heat transfer through the solid cylinder. Generally, as the temperature increases, the steel specific heat and thermal conductivity increase as well. The higher specific heat indicates a greater thermal storage capacity and smaller temperature change for the same energy input. The higher thermal conductivity works to move heat away from the interface at a more rapid rate. Thus, with the combination of the higher specific heat and thermal conductivity, the temperatures at the interface are expected to be lower than those if no temperature variation in the thermal properties were allowed. For the remainder of this study, the temperature varying properties listed for the stainless steel in the material properties section are implemented.

Specific data correlating to the published data at the final mesh and final solid material property demonstrate the current modeling method is valid. First, the macroscopic conditions observed in the contour plots of the pressure and temperature can be compared. With the Mach Number of 6.46 at the hypersonic flow, the flow distribution is as expected with a shock developing upstream of the cylinder due to the presence of the finite cylinder in the high speed flow. The shock structure near the cylinder in the current study as seen in Figure 7 follows that shown in the shadowgraphs of Wieting and in the contour plots of Zhao and Murty including distance ahead of the cylinder where the shock front forms at about 1/6 diameters and the angle the bow shock makes with the direction of the incoming flow to the sides of the cylinder. A high pressure zone is found just ahead of the cylinder and the sharp distinction between this high pressure area and the low pressure zone further upstream can be clearly seen in the pressure contours in Figure 7. The peak system pressure value of 35300Pa is in good agreement with the 33000Pa reported by Zhao.

The temperature field in the gas follows similar trends to those for the pressure field as the two properties are interdependent. The temperature contour from the current model at 5s is presented in Figure 8. The shock development in response to the presence of the cylinder leads to the high gas temperature downstream of the shock, at which the gas properties change abruptly. This high temperature zone ceases upon contact with the solid cylinder, at the cylinder stagnation zone as seen in Figure 8. The peak temperature is 1996K in the current model, closely following the 2000K reported by Zhao. The temperature gradients in the gas and in the solid are highest near the forward facing direct impingement area and decrease moving along the cylinder surface to the rear portion of the cylinder in the wake region. Hence, the conjugate heat transfer to the cylinder caused by such a gas temperature distribution and flow field generates higher heat fluxes and higher temperatures at the front of the cylinder and the lower temperatures to the rear in the wake of the cylinder. The temperature distribution is multi-dimensional. Hence, the macroscopic pressure and temperature field data match well with published data.

Figure 7. Pressure distribution for baseline cylinder case (in Pa).

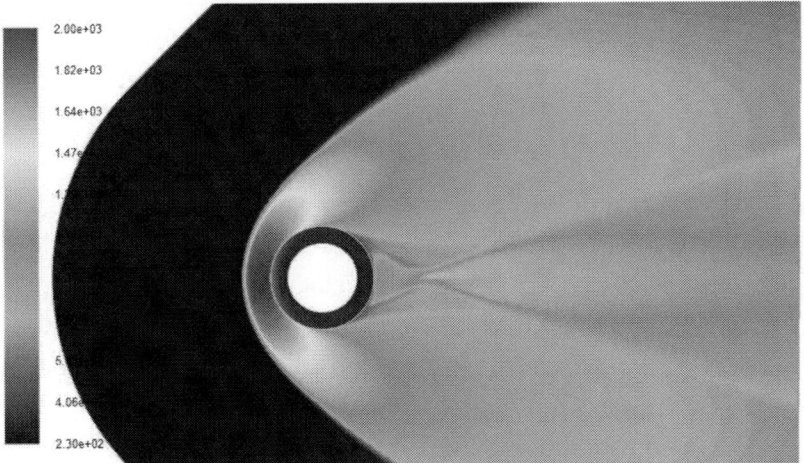

Figure 8. Temperature distribution for baseline cylinder case (in K).

The local normalized heat flux distribution for the final mesh used shown in Figure 5 emphasize the good agreement between the model and published pressure and heat transfer. Thus, the local distribution of the pressure and heat flux indicates the model is able to properly capture the localized phenomena.

# RESULTS

With good comparison to published data and a baseline for the modeling methods and settings established, the two main parts of this study are then conducted. First, the major modeling methods are altered as indicated in Table 1 and the impact on the pressure, temperature, and heat flux at the fluid-solid interface are investigated. Then, for the remainder of the study, the conditions that develop when the cylinder moves through the fluid are studied for three different flow velocities. The results provide information on the modeling methods that may be more appropriate for modeling conjugate heat transfer and further details on the temperature field produced by the high speed compressible flow and the variation that occurs with flow speed.

**Modeling Method Variations**

A series of studies is conducted for the cylinder system in Figure 1 to vary the modeling method or parameter as indicated in Table 1. Contour plots of the pressure, temperature, and velocity fields are compared. The local pressure, temperature, and heat flux around the outer surface of the cylinder are also compared at 0.40s from the start of each simulation. In these studies, a timestep of 1.0e-07s is applied for the baseline model. For comparison purposes, the contours at the same time and view for the baseline case are provided in Figure 9.

The contour plots for the baseline plot clearly show the shock that develops at the forward facing surface of the cylinder, with a delineated region of high pressure and temperature. The stagnation zone at the front of the cylinder, or zero degree orientation relative to the incoming flow can also be clearly observed in the plots in Figure 9. The lower pressures and temperatures to the rear or wake region of the cylinder indicates how the thermal loading on the cylinder is highly position dependent.

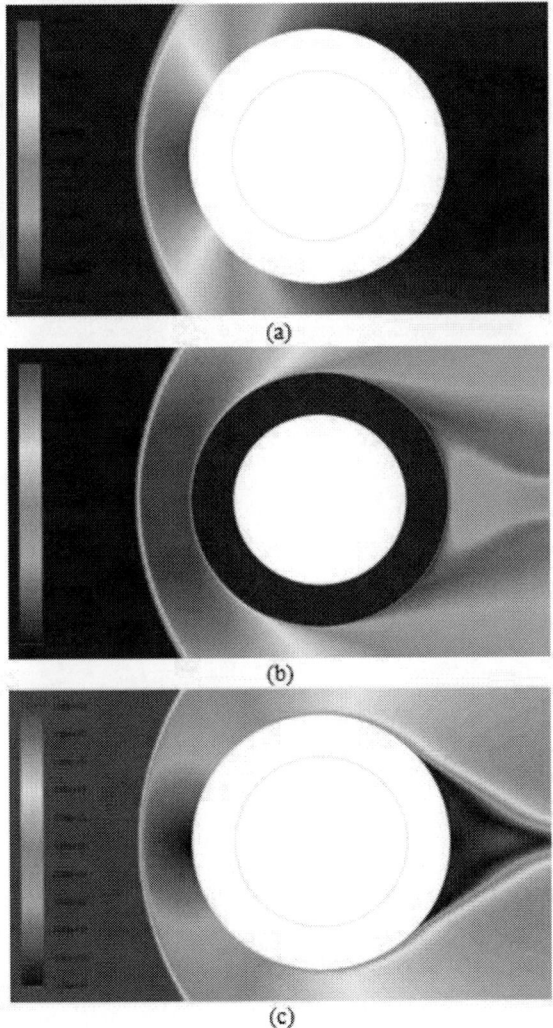

Figure 9. Baseline model contours at 0.4s: (a) Pressure in Pa; (b) Temperature in K; (c) Velocity in m/s.

92                          *Laurie A. Florio*

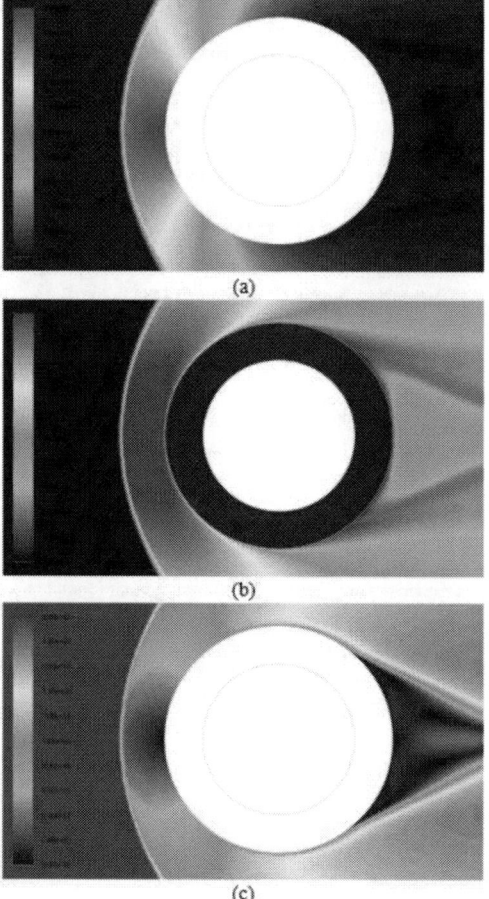

Figure 10. Case A: First order time discretization model contours at 0.4s: (a) Pressure in Pa; (b) Temperature in K; (c) Velocity in m/s.

Now, the effects of altering the computational modeling techniques on the flow and thermal conditions at the 0.40s time are described and the results compared to those for the baseline model.

## *Case A: Time Discretization Method*

The first modeling selection varied is the method by which time derivatives are discretized in the differential equations. For this Case A, the time discretization was switched to the first order discretization for time from second order time discretization in the baseline modeling method. With a given variable, ϕ, and where n+1 is the new time and n is the current time, for timestep Δt, the second order time discretization is given by (Anderson 1995):

$$\frac{3\phi^{n+1} - 4\phi^n + \phi^{n-1}}{2\Delta t} = F(\phi^{n+1}) \qquad \text{Eq. (7)}$$

While the first order time discretization is given by (Anderson 1995):

$$\frac{\phi^{n+1} - \phi^n}{\Delta t} = F(\phi^{n+1}) \qquad \text{Eq. (8)}$$

Figure 10 shows the temperature and pressure contours for this first order transient case and Figure 11 shows the pressure, temperature, and heat flux values as a function of time along with the data for the baseline modeling configuration. Clearly from plots of the pressure, temperature, and heat flux around the cylinder interface surface at time of 0.4s, the first order time discretization does not have a significant impact on the local data for the conditions studied. The time step used in these studies may be adequate to accurately resolve the flow and thermal conditions. Hence, for a particular problem, in order to ensure sufficient resolution of the results of interest, a second order time discretization model should be tested and results compared to those for a first order discretization.

## *Case B: Timestep*

In addition to the time discretization method, the timestep size can also affect the model predictions for the flow and thermal conditions in the fluid and the resulting conjugate heat transfer into the cylinder material. For this Case B, the timestep is doubled from that in the baseline model, with all other modeling methods and parameters held the same as the baseline methods/values. Hence, the timestep for this case was raised to 2.0e-07s from 1.0e-07s. The contour plots for Case B are in Figure 12.

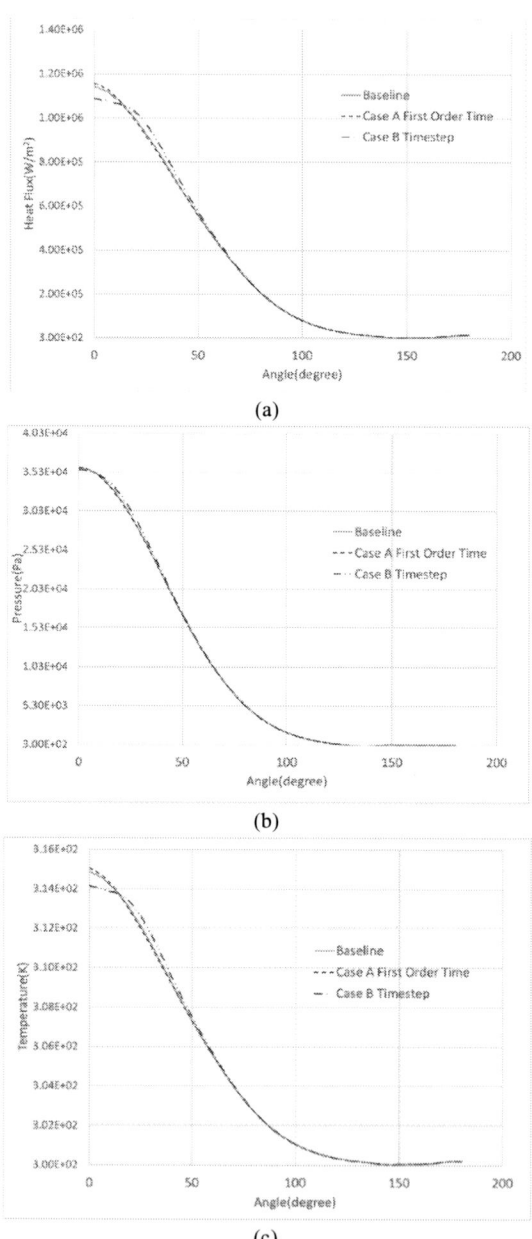

Figure 11. Time related modeling method variation, comparison of local distributions at outer cylinder surface: (a) Heat Flux in W/m$^2$; (b) Pressure in Pa; (c) Temperature in K.

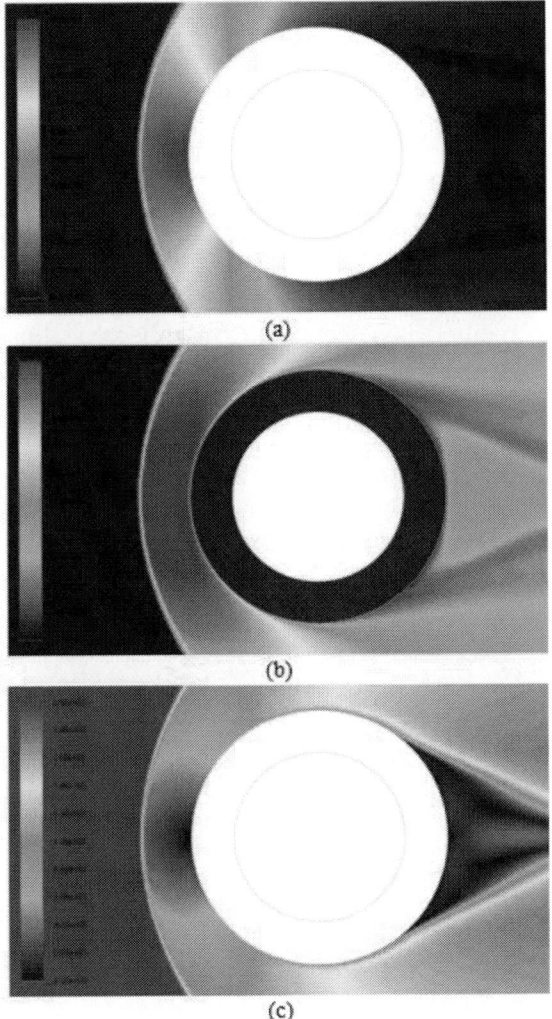

Figure 12. Case B: Timestep changed from 1.0e-07s to 2.0e-07s model contours at 0.4s: (a) Pressure in Pa; (b) Temperature in K; (c) Velocity in m/s.

From the contour plots, the basic features and peak values of the temperature and pressure data and the flow distribution closely follow those of the baseline configuration. Some details in the flow in the stagnation region and in the wake region show more differences from the baseline than the results for the first order discretization with time from Case A. The local quantitative data for the interface pressure, temperature, and heat flux are shown in Figure 11. The heat flux results in Figure 11a show a lower heat flux and temperature near the stagnation zone for the higher timestep case, though the flux curve and temperature curve meet the data for the first and second order time discretization cases with the smaller time step as the angle position towards the rear of the cylinder increases. The sensitivity of the heat flux and temperature to the change in the time step is greater than the sensitivity of the pressure to this same change as seen in comparing Figures 11a and c to Figure 11b. Hence, since an implicit solution method is selected for time discretization in this work, investigating the effect of the timestep size on the model results of interest is needed to check that the solution is independent of the timestep.

### *Case C: Upwinding*

In Case C, a First Order Upwinding discretization method is used instead of the Second Order Upwinding selected for the baseline model. Recall in First Order Upwinding, the value of a variable $\phi$ at the face f of a computational cell, $\phi_f$, is set to the value at the cell centroid on the computational cell that is upwind of the face. This assures the proper information is "carried" or transported into the computational cell. For example, for a 1-D flow with cell 1(C,1) on the left of a boundary and cell 2 (C,2) on the right of the boundary, the face value specified under the different flow conditions is listed below (Anderson 1995):

$$Flow\ left\ to\ right\ \phi_f = \phi_{C,1}$$
$$Flow\ right\ to\ left\ \phi_f = \phi_{C,2} \qquad (Eq.9)$$

For a Second Order Upwinding formulation, the value of a variable $\phi$ at face f, $\phi_f$, is approximated by the cell centroid value $\phi$ and gradient value in the computational cell that is upwind of the face. For the same example of one directional flow with cell 1 on the left and cell 2 on the right, the value of $\phi$ at the cell face is given by the expressions below where **r** is the vector from the upwind cell centroid to the centroid of face f. (Anderson 1995):

*Flow left to right* $\phi = \phi_{C,1}, \nabla\phi = \nabla\phi_{C,1}$
*Flow right to left* $\phi = \phi_{C,2}, \nabla\phi = \nabla\phi_{C,2}$
*Then,* $\phi_f = \phi + \nabla\phi \cdot r$  Eq.(10)

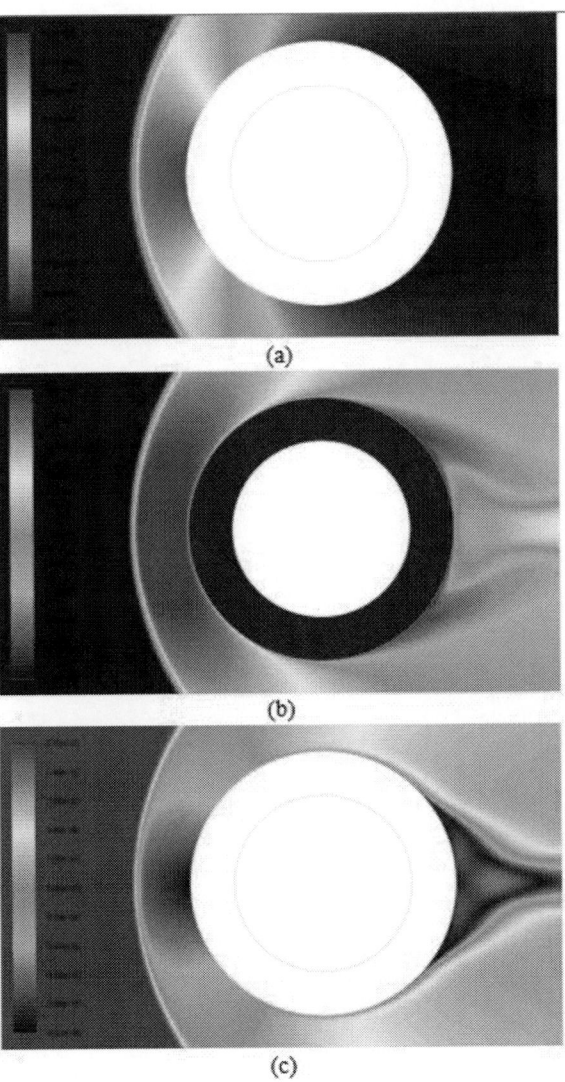

Figure 13. Case C: First order upwinding model contours at 0.4s: (a) Pressure in Pa; (b) Temperature in K; (c) Velocity in m/s.

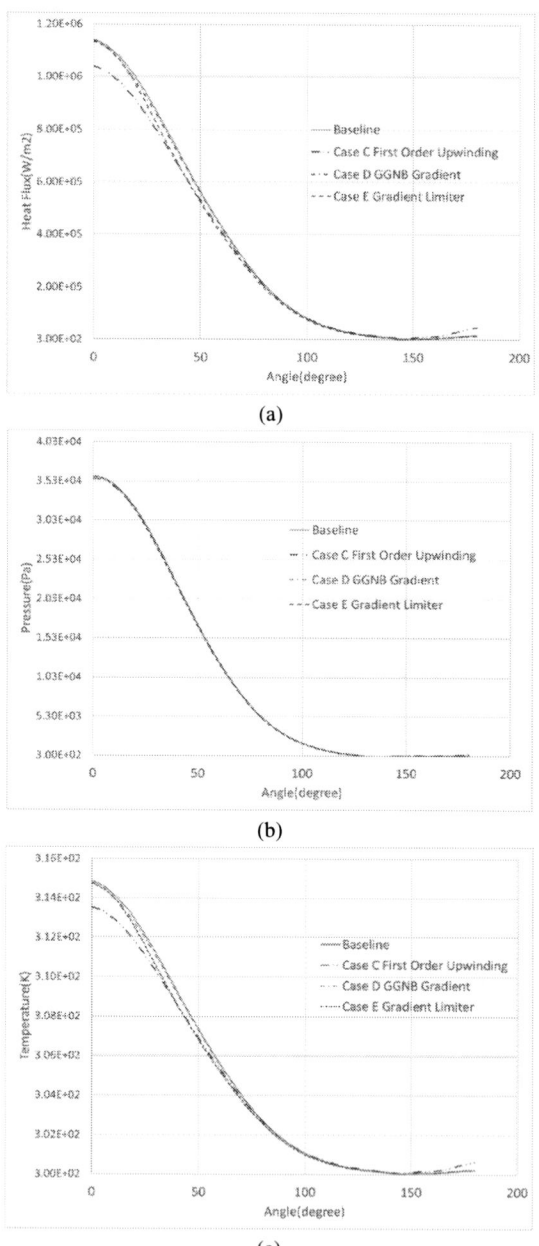

Figure 14. Discretization related modeling method variation, comparison of local distributions at outer cylinder surface: (a) Heat Flux in W/m$^2$; (b) Pressure in Pa; (c) Temperature in K.

The results for the two different upwinding schemes can be compared in the contour plots in Figures 9 and 13 and the local data in Figure 14. The contour plots for the First Order Case in Figure 13 show some noticeable differences in the flow and temperature field most distinctly in the wake region. For the local data along the circumference of cylinder, however, the influence of the upwinding scheme can be seen. Similar to the timestep change, the pressure distribution for the First Order Upwinding exhibits little variation from the results for the Second Order Upwinding in the baseline case. The heat flux and temperature, similar to the higher timestep case, exhibit lower values near the stagnation location at the angular position of zero compared with the Second Order, baseline model results. With the high temperature gradients near the cylinder surface, the accuracy offered by the Second Order Upwinding better resolves the heat flux at the boundary and so the heat conduction through the solid cylinder.

### *Case D: Gradient Calculations*

The results of the application of two techniques of calculating the cell centroid gradient are also examined in the studies conducted. The Least Square Cell Based Gradient method was used in the baseline case. For Case D, the variation case, the Green-Gauss Node Based method are applied. Both methods evaluate the gradient at the cell centroid, c, $(\nabla \phi)_c$, using the values of $\phi$ at the face as in the expression in Equation 11. The means of evaluating the face values, $\phi_f$, differ.

$$(\nabla \phi)_c = \frac{1}{VOL} \sum_f^{Nf} \phi_f A_f \qquad \text{(Eq.11)}$$

For the Green-Gauss Node Based gradient, the face values in Eq. 11 are based on an average of nodal values. The nodal values are estimated by the cell values weighted by the distance from the node to the cell centroid. The Least Square Cell Based gradient utilizes the neighboring cell centroid values to estimate the face values, but uses a least squares method. Further details can be found from (Mishriky and Walsh 2017).

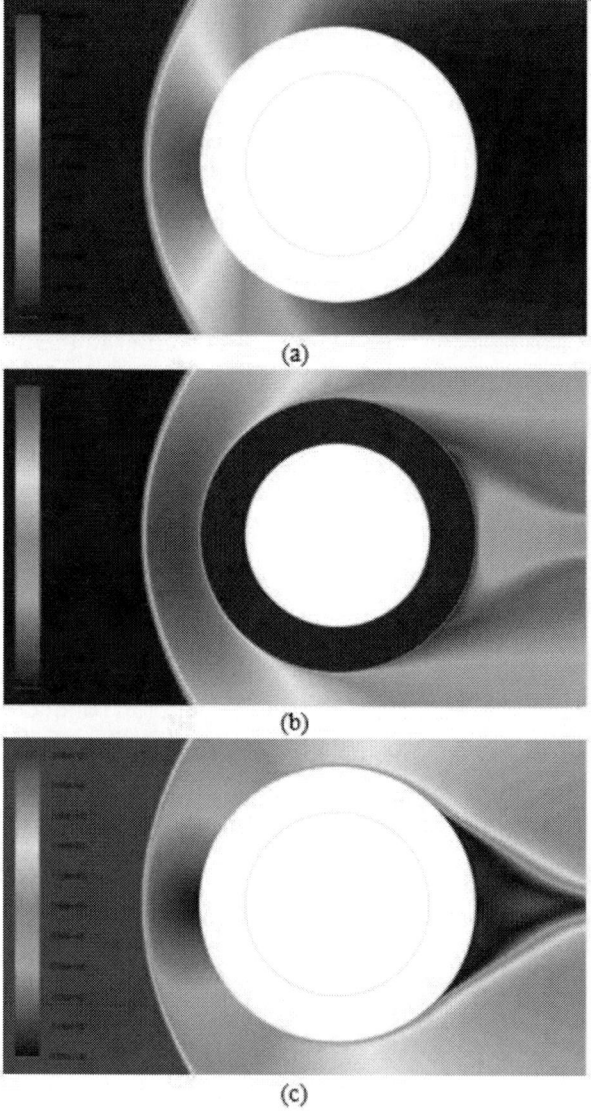

Figure 15. Case D: Green Gauss Node Based Gradient model contours at 0.4s: (a) Pressure in Pa; (b) Temperature in K; (c) Velocity in m/s.

The influence of the gradient method on the predicted flow and thermal conditions can be found by comparing the conditions in the contour plots for the baseline in Figure 9 and the contours for Case D in Figure 15 as well as the local data in Figure 14. The contour plots of the pressure, temperature, and velocity for the Green Gauss Node Based Gradient show no noticeable difference to the Baseline Case flow field and temperature field. For the local heat flux, pressure and temperature about the outer surface of the cylinder, the data in Figure 14, the Case D values are all slightly lower than those for the Baseline Case, but differ by at most less than one percent. Hence, for this particular system, the gradient method has little influence on the results and this finding is consistent with the literature (Mishriky and Walsh 2017).

*Case E: Gradient Limiter*

The gradient limiter is meant to limit oscillations in the solution that appear near shocks or discontinuities in the solution. The limiters "clip" overshoots in the calculation of the face values in the Second Order Upwinding by limiting the magnitude of the gradient in Equation 10. The Standard limiter is based on a minimum value and modulus function (Bartch 1989). A multidimensional limiter is employed instead for Case E, allowing for unequal "clipping" in the different coordinate directions (S. C. Kim 2003).

The data produced by Case E and the Baseline model can be compared to gain knowledge about the influence of the gradient limiter type. The solution contour plots in Figure 16 depict the overall Case E simulation flow and temperature fields with the gradient limiter and show little difference from those results for the Baseline Case. For the local data presented in Figure 14, once again, the limiter has no significant influence on the pressure level distribution around the circumference of the cylinder. For the heat flux and the temperature levels, the change to the Multidimensional limiter has some effect on the flow results. The peak heat flux and temperatures are comparable, but then the values drop more rapidly moving around the cylinder. The maximum percent difference is about 7%. Traveling around to the rear of the cylinder all curves unify and follow the same distribution. Hence, for a particular application, the effects of varying the limiter options on the numerical the results should be examined and the model results compared to known solutions or results.

102  Laurie A. Florio

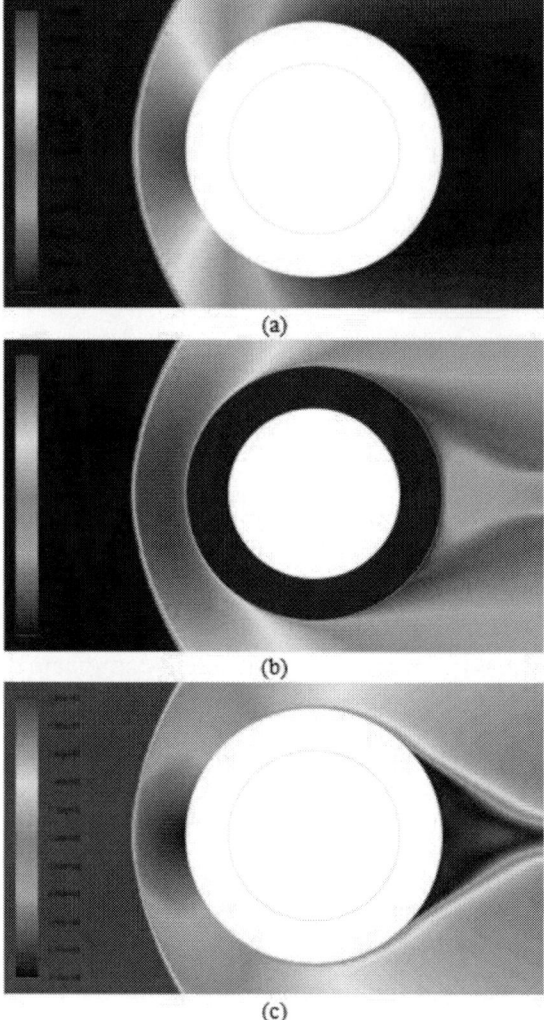

Figure 16. Case E: Gradient Limiter model contours at 0.4s: (a) Pressure in Pa; (b) Temperature in K; (c) Velocity in m/s.

## Case F: Compressibility Effects with κ–ω Model

The compressibility effects option introduces a factor to account for the compressibility of the flow in the turbulence model. The parameter selection is based on experimental data and likely requires tuning for a particular system. Case F turns the compressibility effects option on with default values (Wilcox 2006).

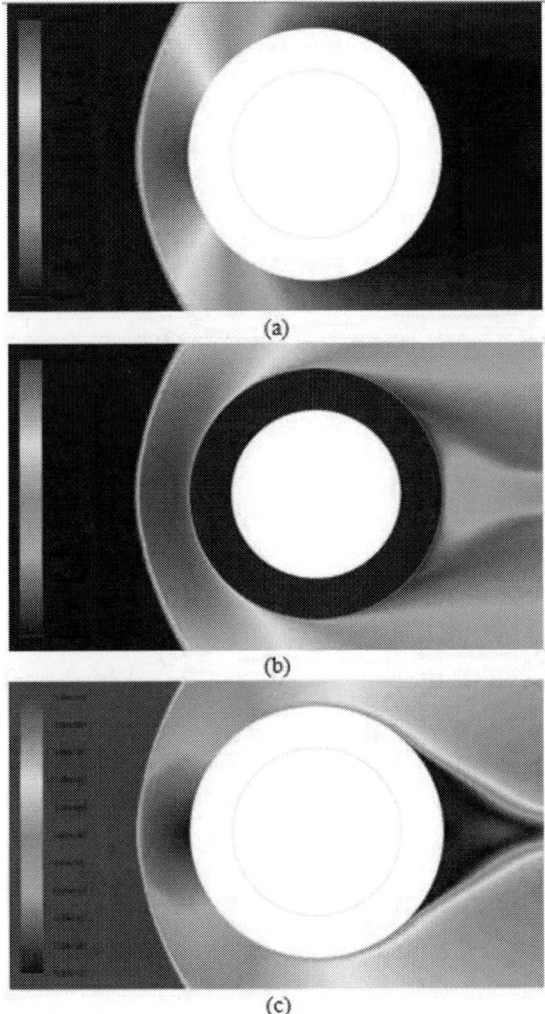

Figure 17. Case F: Compressibility Effects on model contours at 0.4s: (a) Pressure in Pa; (b) Temperature in K; (c) Velocity in m/s.

The comparison of the results with and without the compressibility effects provides insight into the solution values that are affected by this option in the turbulence model. The overall or bulk changes in the flow field with this compressibility effects model can be seen in Figure 17 compared to the Baseline results in Figure 9. Only some slight variations in the fields compared to those for the Baseline case are found in the results for Case F with the compressibility effects on. As in the other cases, the most noticeable changes occur in the wake region of the cylinder. Also, following trends in the other parameters and settings, as shown in Figure 18, the pressure distribution around the cylinder at 0.4s is not highly affected by the application of the compressibility effects. However, the thermal conditions at the interface between the flow and the cylinder, and thus the heat conduction through the cylinder do show differences from those of the Baseline Case. The temperature and heat flux at the stagnation point is a maximum of two percent different for the Case F and Baseline models. The difference decreases proceeding around the cylinder, with the curves for these two cases nearly identical after about 80 degrees. Because the parameter used for the compressibility effects is experimental, it is not typically recommended that this model be used for a general system.

### *Case G: Standard $\kappa$–$\omega$- Turbulence Model*

The $\kappa$–$\omega$-SST turbulence model was selected for the baseline method since it was developed to better capture separated flow compared with the Standard $\kappa$–$\omega$ model. The standard model typically predicts a higher value of the heat flow and higher temperatures than the $\kappa$–$\omega$ SST model (Wilcox 2006). Whether such a difference in the heat flow occurs with the current system can be seen by comparing the results for Case G which implements the Standard $\kappa$–$\omega$ turbulence model and the Baseline model set-up.

Examination of the results for Case G and the Baseline clearly show the standard model alters the predicted flow and thus the conduction in the cylinder. With the Standard $\kappa$–$\omega$ turbulence model (Figure 19), the shape and values of the flow and temperature fields, particularly the pressure and temperature in the stagnation area and the temperature field in the wake, are affected by the turbulence model. Locally, a 52% increase in the peak in the stagnation heat flux and a 5% increase in the peak temperature with the Standard $\kappa$–$\omega$ turbulence model option can be observed in Figure 18. The differences are maintained for a significant portion of the circumference of the cylinder until about 70 degrees, though the differences diminish moving towards the rear of the cylinder.

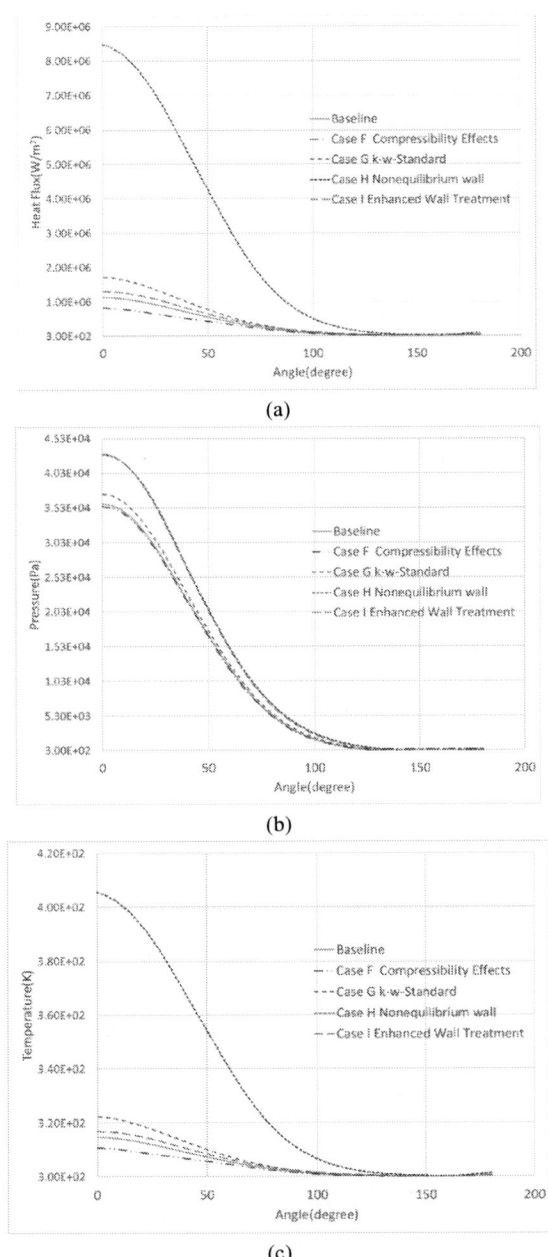

Figure 18. Turbulence related modeling method variation, comparison of local distributions at outer cylinder surface: (a) Heat Flux in W/m$^2$; (b) Pressure in Pa; (c) Temperature in K.

These findings are consistent with the comments of Wilcox (Wilcox 2006) that the standard model does not treat heat transfer with separated flows properly and results in larger heat fluxes than other models such as the SST. If the thermal conditions at the wall and cylinder are of interest, then this model should most likely not be used.

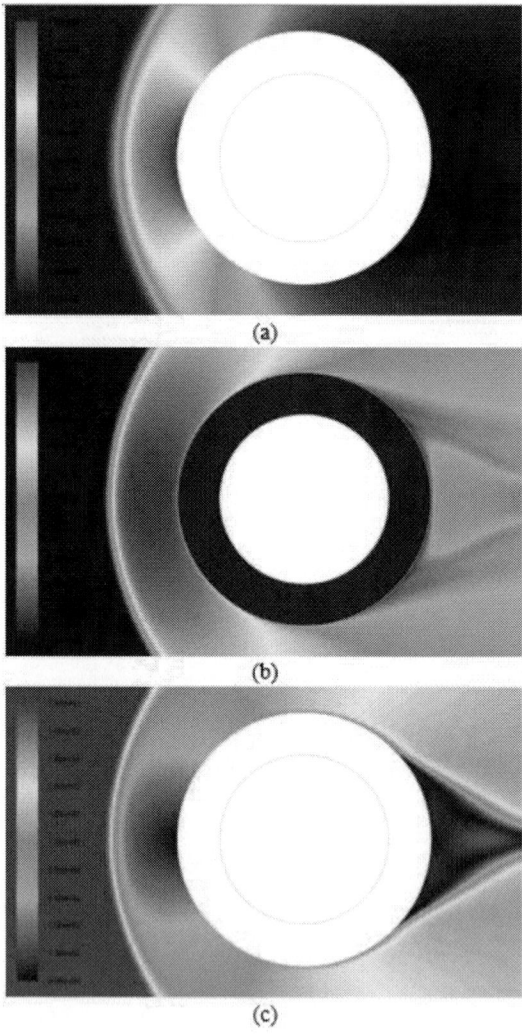

Figure 19. Case G: Standard κ–ω turbulence model contours at 0.4s: (a) Pressure in Pa; (b) Temperature in K; (c) Velocity in m/s.

### Case H: Non-Equilibrium Wall Treatment, κ–ε Turbulence Model.

In general, since the κ–ε turbulence model utilizes the log law of the wall and does not resolve the flow and thermal conditions to the wall, the model is not recommended for estimating the heat transfer to the wall for a conjugate heat transfer problem like the one considered in this chapter. However, some modifications were made to the standard κ–ε turbulence model. The κ–ε-non-equilibrium-wall-treatment is a specialized wall treatment method particularly for separated flows and reattachment since it accounts for the pressure gradient variation neglected in the wall function. Friction coefficients and heat transfer are better approximated using this approach (S. E. a. Kim 1995). Case H implements this non-equilibrium wall treatment with the κ–ε turbulence model.

The predictions from this κ–ε model with non-equilibrium wall treatment may be compared to the predictions with the Baseline model with the κ–ω SST turbulence model. The Case H results have the greatest deviation from the Baseline for any of the models considered in this chapter. Figure 20, the system contour plots, clearly demonstrates this trend. The stagnation zone shows a much smaller region of the highest pressure and a larger, more diffuse region of the highest temperatures. In the wake, the temperatures are higher and the distinctive low velocity zone is missing. The peak temperature is 28% higher than that for the Baseline with the rise from the initial temperature about seven times that for the Baseline (Figure 18). Because of the large difference in temperatures and heat flux values with this model compared to the κ–ω SST, which produced a good comparison to the literature, this non-equilibrium wall treatment is not recommended for studying a conjugate heat transfer problem like that in the current chapter.

### Case I: Enhanced Wall Treatment – κ–ε- Turbulence Model

The enhanced-wall-treatment is used to remove the requirement for the κ–ε type turbulence model that the near wall mesh either be resolved in the viscous sublayer or begin outside of this viscous sublayer in the fully turbulent flow area. Thus, the treatment allows for reduced dependency on the near wall mesh characteristics. The mesh near the wall can have a $y^+$ near 1 or $y^+$ larger than 1 without sacrificing accuracy (Kader 1981).

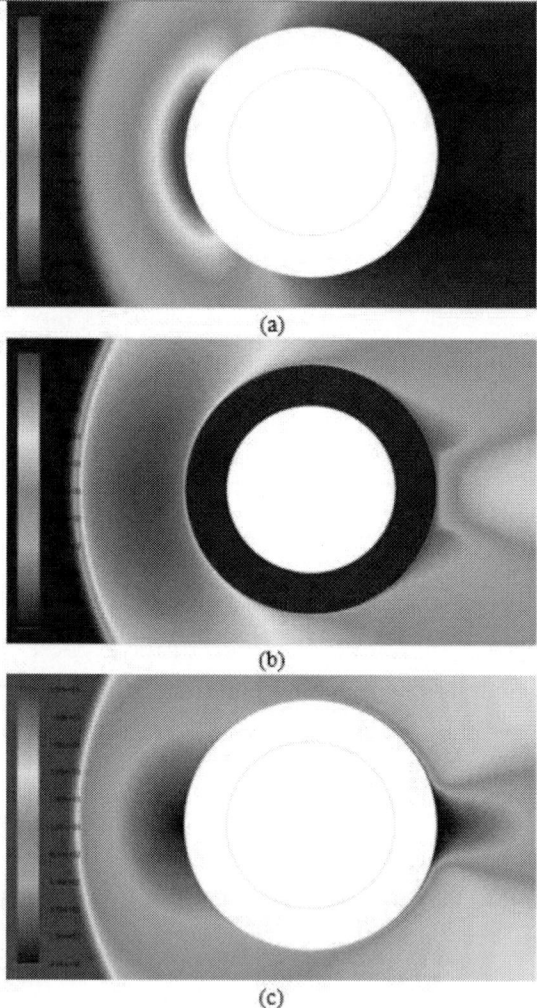

Figure 20. Case H: Non-equilibrium $\kappa$–$\varepsilon$, turbulence model contours at 0.4s: (a) Pressure in Pa; (b) Temperature in K; (c) Velocity in m/s.

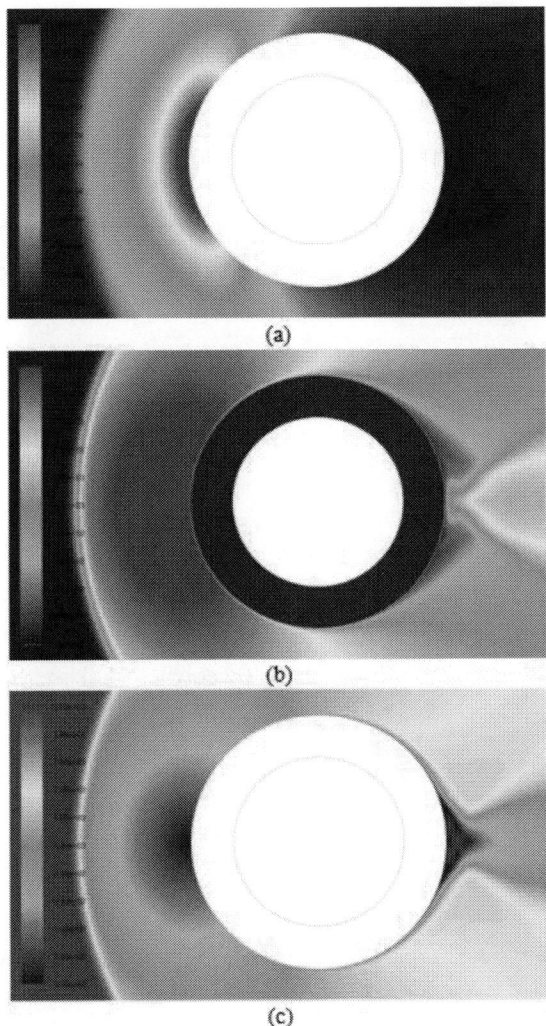

Figure 21. Case I: Enhanced wall treatment, turbulence model contours at 0.4s: (a) Pressure in Pa; (b) Temperature in K; (c) Velocity in m/s.

The impact of this enhanced wall treatment model on the conditions predicted can be found by comparing the Case I results with those of the Baseline model. A smaller region of the highest pressures is found in the stagnation zone with a larger high temperature zone (Figure 21), much like the non-equilibrium wall treatment of Case H. An elevated temperature region is found in the wake region of the cylinder and the wake region low velocity zone is significantly reduced in size, with some higher velocity flow moving around the sides of the cylinder. On the local level, as seen in Figure 18, the heat flux and temperature values at the wall are only about 10% different at the stagnation point compared to the Baseline conditions, despite the significant differences in the general flow patterns. The local pressure closely follows that from the non-equilibrium wall treatment (Case H) with about a 20% higher stagnation pressure than the baseline case. Given the significant differences in the general flow field shape and features with the model, for the conditions of interest, this model does not seem appropriate for capturing the thermal conditions resulting from the high speed flow over a cylinder.

## Moving Cylinder Modeling Method

The final set of models involves now moving the cylinder through the fluid at three different velocities using three different methods of moving the mesh to account for the cylinder motion through the flow. The results at the three velocities: subsonic, near sonic, and supersonic velocity regimes, are compared so that the moving mesh method best suited for a particular flow regime can be selected. In these models, the baseline modeling methods and parameters are used. The second order discretization is changed to first order discretization and a time step of 1.0e-06 seconds is used across all of the methods. Comparisons of the results discussed and presented are for the same displacement of the cylinder through the fluid at the different velocities.

For each of the cylinder velocity values, the results of the simulations using the three moving mesh techniques are reviewed.

### *Velocity = 250 m/s*

For this speed, the flow conditions are subsonic. Since the cylinder velocity is lower than the speed of sound, the shock pattern found for the fixed cylinder cases does not form for this case and a wide region of elevated temperature develops ahead of the cylinder. The lower cylinder and resulting fluid speed and the resulting lower level of gas heating produces a lower level input of energy into the solid cylinder.

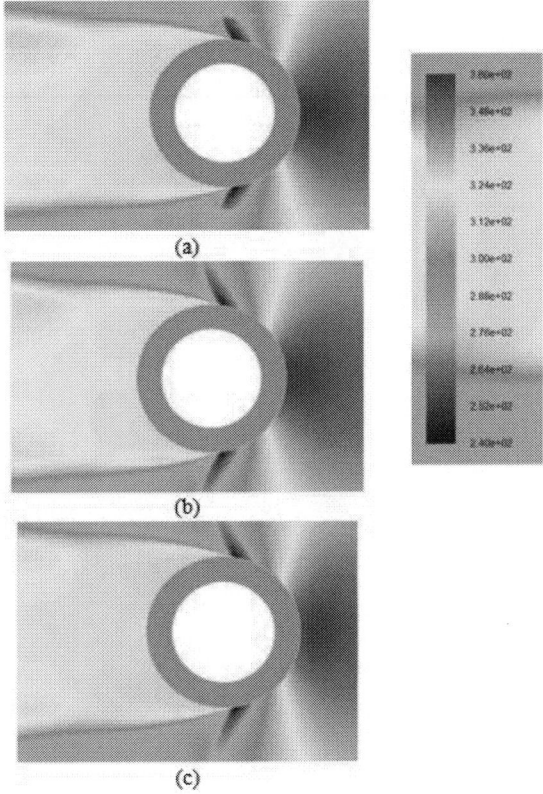

Figure 22. Temperature Contours (K) for case with Velocity of 250m/s: (a) Sliding; (b) Remeshing; (c) Overset.

The results for the three moving mesh methods at the same displacement are plotted in Figure 22 and show little difference in the predicted temperature field. Examining the local data in Figure 23, the pressures, temperatures, and heat flux at the fluid-cylinder interface are typically higher for the remeshing method, then lower for the sliding mesh method, and lowest for the overset method. Overall the difference is under 1% due to the low level of the temperature rise that occurs for this fluid velocity.

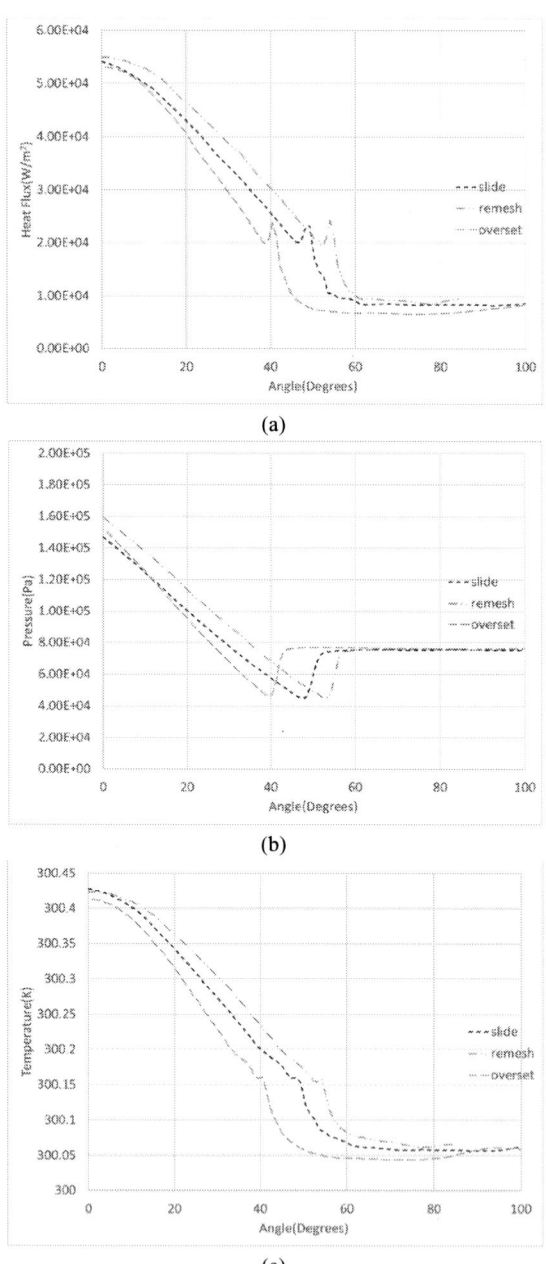

Figure 23. Local data for case with Velocity of 250m/s: (a) Heat Flux (W/m$^2$); (b) Pressure (Pa); (c) Temperature (K).

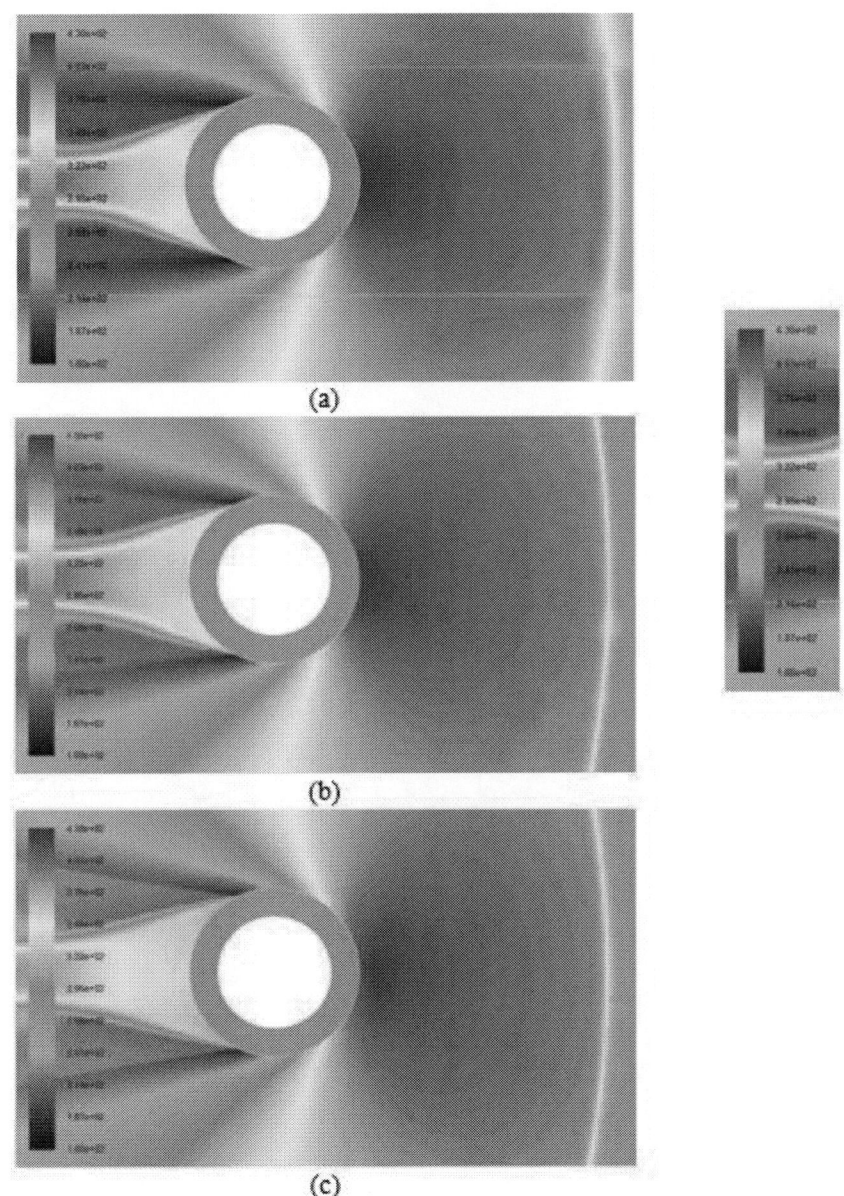

Figure 24. Temperature Contours (K) for case with Velocity of 500m/s: (a) Sliding; (b) Remeshing; (c) Overset.

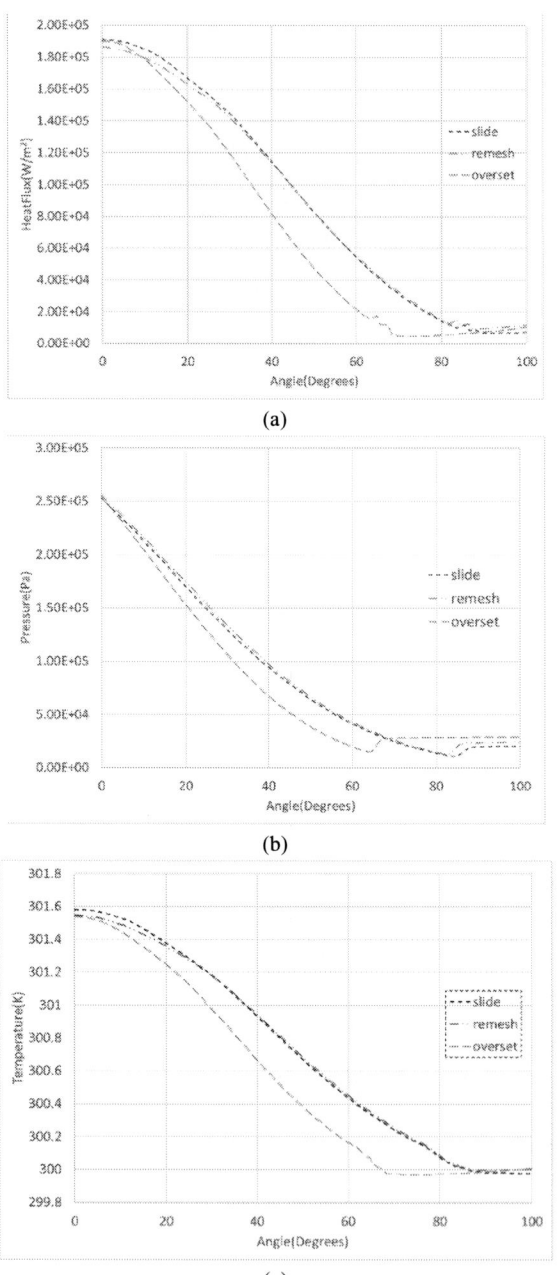

Figure 25. Local data for case with Velocity of 500m/s: (a) Heat Flux (W/m$^2$); (b) Pressure (Pa); (c) Temperature (K).

With the need for interpolation to the background mesh and no gradient refinement utilized for the overset model, the lower values for the overset model follow expected trends. The sliding approach also requires interpolation across the interfaces where the sliding is occurring, though the interfaces are not in the immediate proximity of the cylinder interface. Perhaps the interpolation contributes to the lower values of the solution field parameters for the sliding mesh motion compared to the remeshing. The difference in the mesh type and therefore mesh size may also play a role and require further investigation. Another observation is the differences in the results between the three methods are largest for this lower velocity case. Perhaps the longer time to react to the momentum change before the cylinder arrives at the same position with the lower speed plays a role in this trend.

### *Velocity = 500 m/s*

For this velocity, the Mach Number of the flow directly induced by the cylinder motion is greater than one. The differences in the flow pattern around the cylinder compared to the results for the 250 m/s cylinder speed are apparent as seen by the temperature plots in Figure 24. The differences in the temperature field in the wake region can be seen in the contour plots. Again, the lower temperature levels are found for the overset mesh method. These lower temperatures are then likely due to the lack of the use of a dynamic mesh refinement on the fixed background mesh to better capture the conditions near the cylinder. The local data for these three cases are provided in Figure 25. The overset method continues to show the lowest temperature, pressure, and heat flux values amongst the three moving mesh methods. With the shorter time period to reach the same displacement, the results for the sliding mesh and remeshing cases are comparable for this case. In addition, the values or levels of all of the measured temperature, pressure, and heat flux parameters increase with the cylinder velocity. The peak temperature and heat flux are about four times greater than those with the 250m/s velocity and the peak pressure is about double that for the 250 m/s case. The models for this case could be run for a longer time to ensure the good comparison between the remeshing method and sliding method are maintained, or whether momentum effects eventually cause differences in the results. Again, further mesh sensitivity tests could be conducted to determine if the differences in the mesh type between the remeshing and the sliding methods have any influence on the results.

### *Velocity = 1000 m/s*

Finally, the case where the cylinder is accelerated to 1000 m/s is explored. With this velocity, the flow induced directly by the motion of the cylinder is at a Mach Number near 3. The contour plots in Figure 26 show the more diffusive temperature pattern with the overset. The sharp shock for the overset is smeared out, likely due to the interpolation to the coarser base background mesh. This more diffusive temperature pattern is consistent with the local data results. The remeshing and the sliding methods produce more similar contour patterns. The local data in Figure 27 shows the higher fluid induced pressures and temperatures that develop, creating higher heat fluxes moving into the cylinder. The peak stagnation pressure is about an order of magnitude higher than the 500 m/s case, the temperature rise about 20 times greater and the heat flux also an order of magnitude higher. The greater energy carried with the higher speed flow is converted to pressure and thermal energy as the gas flow is slowed by the presence of the cylinder and the higher energy level is responsible for these differences. Again, these plots, as with those for the 500m/s case, show the overset method data at lower levels than the sliding and remeshing mesh motion methods at this speed.

The sliding mesh and remeshing techniques, however, follow each other more closely than for the 500 m/s case. This trend is likely due to the faster speed of the cylinder and shorter time requied to get to this same position. The fluid and cylinder have a more limited time to "react" to the motion than for the slower speed cases. For this time and speed, there is little difference in the local data produced by the use of the sliding and overset methods.

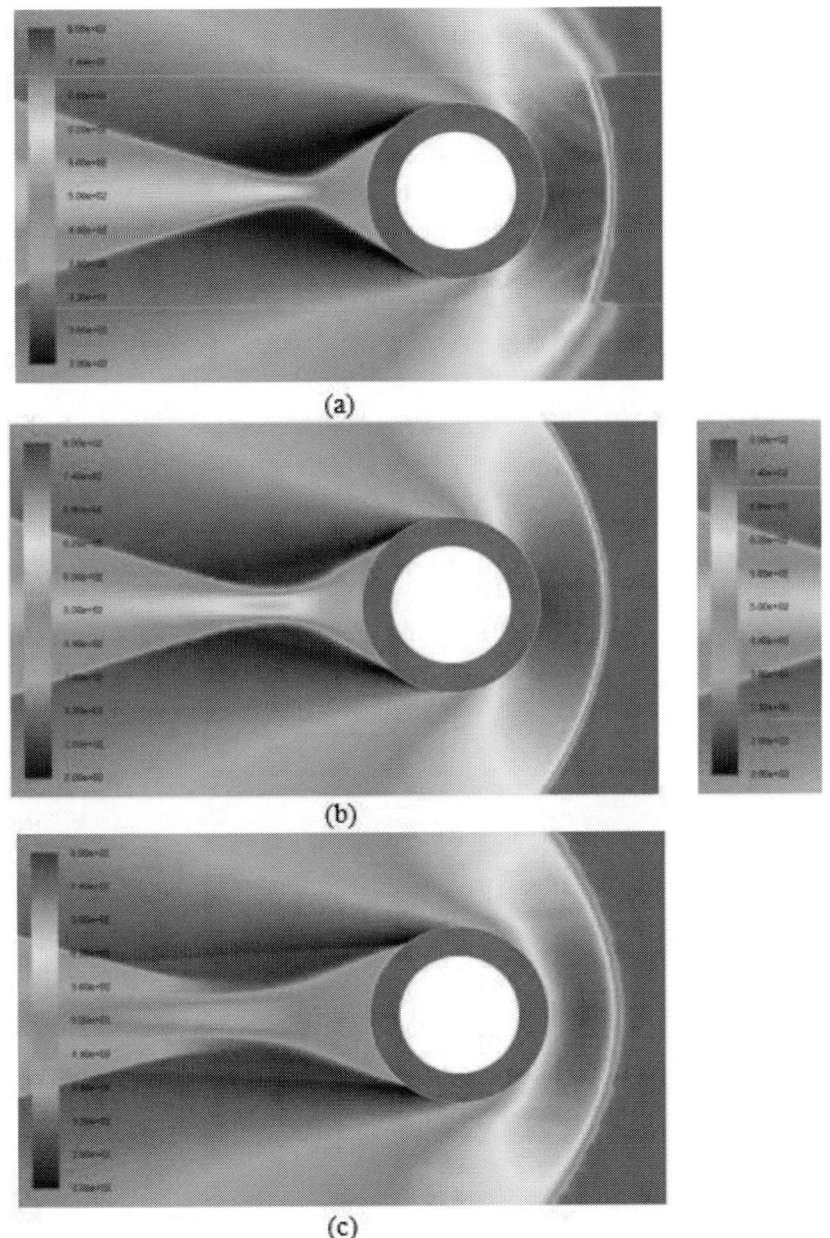

Figure 26. Temperature Contours (K) for case with Velocity of 1000m/s: (a) Sliding; (b) Remeshing; (c) Overset.

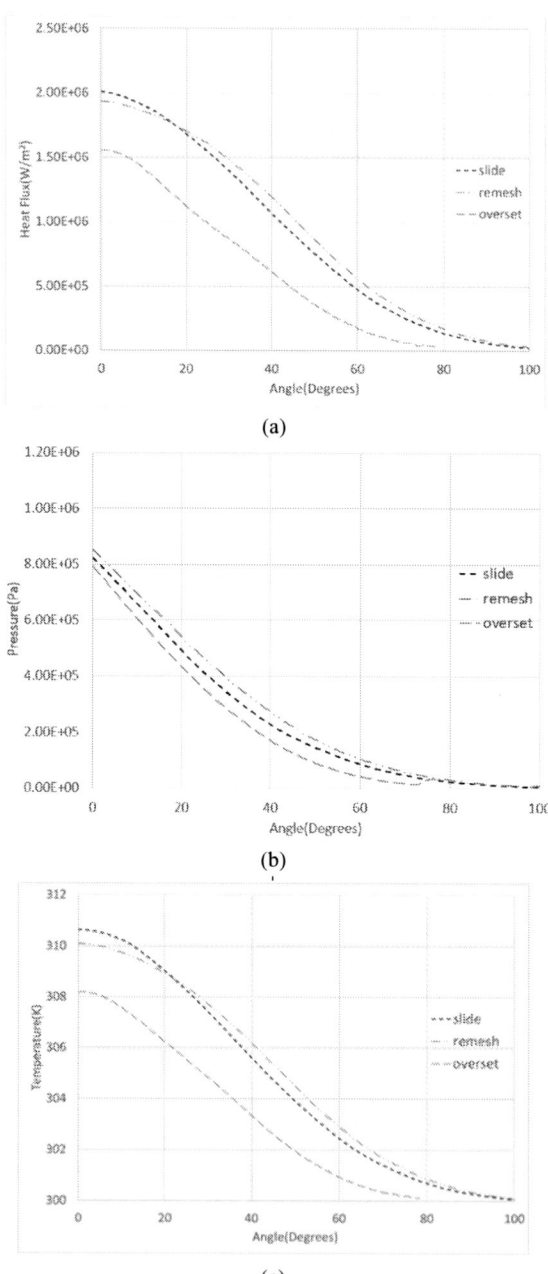

Figure 27. Local data for case with Velocity of 1000m/s: (a) Heat Flux (W/m$^2$); (b) Pressure (Pa); (c) Temperature (K).

## Conclusion

The results of this investigation provide important information regarding the influence of the modeling methods on the temperature field, and the velocity and pressure that develop with a cylinder or other object placed in a high speed compressible flow. For the specific cases and conditions considered, in order to capture the thermal conditions at a fluid-solid interface under high speed flow conditions, the sensitivity to timestep should be examined and the Second order upwinding should be used. A gradient limiter and second order time discretization may be tested to determine if the limiter and the higher order time discretization assist with increased accuracy of solution for the particular problem and necessary output, especially when known published data exists. The First Order Upwinding is not suggested for this type of flow. Changes in these modeling methods or parameters have more effect on the local temperature and heat flux than the pressure, so if thermal conditions are of interest more care should be taken to ensure the results are timestep independent and take into account the necessary phenomena.

For the turbulence related modeling methods, the change in the results brought about by altering the turbulence model options were more significant. However, based on the basic changes in the flow patterns and the local pressure, temperatures, and heat flux conditions, the modifications to the model did not bring about any benefit over the baseline model for capturing the local thermal conditions at the fluid-solid interface in general. The $\kappa$–$\varepsilon$ modifications did not produce realistic flow field and in general had higher temperatures and heat fluxes than those from the $\kappa$–$\omega$ model. The compressibility factor produced changes in the flow distribution as well and noticeable changes in the local temperature and heat flux distributions around the circumference of the cylinder. However, the need to select a parameter based on experimental data limits its applicability. Consistent with the literature, the standard $\kappa$–$\omega$ model produces high heat fluxes for conditions where flow separation occurs. Thus, none of the methods examined improved the modeling of the conjugate heat transfer. More advanced turbulence models could be investigated.

For the conditions considered, the sliding interface and remeshing methods for the cylinder motion through a fluid produce similar pressure, temperature, and heat flux data, particularly at the higher speeds. The overset mesh motion method needs to be tested with a velocity or temperature value or gradient value based dynamic refinement on the background mesh to investigate if this assists in better resolving the flow and thermal conditions since the overset consistently predicted the lowest pressures, temperatures, and velocities for all three of the cylinder speeds. Additional mesh sensitivity studies can be conducted for the sliding mesh and remeshing based methods to investigate if the mesh size or type leads to the differences in the results, particularly at the lower cylinder speeds. Based on the information obtained in the model, the use of the sliding or remeshing methods produces more accurate results.

The modeling approaches selected must take into account the phenomena of interest and must be attuned to accurately capture the specific output of interest for a given investigation. This work has provided insight into the modeling approaches that should be considered for high speed compressible flow conditions with conjugate heat transfer.

## REFERENCES

Anderson, J. D. 1995. *Computational Fluid Dynamics: The basics with applications.* New York: McGraw Hill.

Bartch, T. J. and Jespersen, D. 1989. "The design and application of upwind schemes on unstructured meshes." *AiAA 27$^{th}$ Aerospace Sciences Meeting. Reno, Nevada: AIAA.* AIAA-89-0366.

Kader, B. 1981. "Temperature and concentration profiles in fully developed turbulent boundary layers." *Int. Journal of Heat and Mass Transfer* 24 (9): 1541-1544.

Kim, S., Caraeni, D., Makarov, B. 2003. "A multidimensional linear resconstruction scheme for arbitrary unstructured grids." *AIAA 16$^{th}$ Computational Fluid Dynamics Conference*. Orlando, FL: AIAA.

Kim, S.-E. and Choudhury, D. 1995. "A near-wall treatment using wall functions sensitized to pressure gradient." *ASME FED Separated and Complex Flows* 217.

Mishriky, F., and P. Walsh. 2017. "Towards understanding the influence of the gradient reconstruction methods on unstructured flow simulations." *Transactions of the Canadian Society for mechanical Engineering.*

Murty, Chanda MSR, P. Manna, and Debasis Chakraborty. 2012. "Conjugate heat transfer analysis in high speed flows." *Proc. IMechE Part G: Journal of Aerospace Engineering* 227 (10): 1672-1681.

Patankar, S. V. 1980. *Numerical heat transfer and fluid flow*. Washington, D.C.: Hemisphere Publishing Corporation.

White, F. M. 1998. *Viscous Fluid Flow*. New York: McGraw Hill.

Wieting, A. R. 1987. *"Experimental study of shock waveinterference heating on a cylindrical leading edge,"* NASA TM-100484.

Wilcox, D. C. 2006. *Turbulence modeling for CFD*. Mill Valley, CA: DCW Industries.

Zhao, Xiaoli, Zhenxu Sun, Longsheng Tang, and Gangtie Zheng. 2011. "Coupled flow-thermal-structural analysis of hypersonic aerodynamically heated cylindrical leading edge." *Engineering Applications of Computational Fluid Mechanics* 5 (2): 170-179.

In: Understanding Heat Conduction
Editor: William Kelley
ISBN: 978-1-53619-182-0
© 2021 Nova Science Publishers, Inc.

*Chapter 3*

# ADVANCES IN HEAT CONDUCTION ANALYSIS WITH FUNDAMENTAL SOLUTION BASED FINITE ELEMENT METHODS

## *Qing-Hua Qin*[*]
Department of Mechanics,
Tianjin University, Tianjin, China

### ABSTRACT

This chapter presents an overview of the fundamental solution (FS) based finite element method (FEM) and its application in heat conduction problems. First, basic formulations of FS-FEM are presented, such as the nonconforming intra-element field, auxiliary conforming frame field, modified variational principle, and stiffness equation. Then, the FS-FE formulation for heat conduction problems in cellular solids with circular holes, functionally graded materials, and natural-hemp-fiber-filled cement composites are described. With this method, a linear combination of the fundamental solution at different points is actually used to approximate the field variables within the element. Meanwhile, the independent frame field defined along the elemental boundary and the modified variational

---

[*] Corresponding Author's E-mail: qhqin@tju.edu.cn.

functional are employed to guarantee inter-element continuity as well as to generate the final stiffness equation and establish the linkage between the boundary frame field and the internal field in the element. Finally, a brief summary of the approach is provided and future trends in this field are identified.

**Keywords**: fundamental solution, finite element method, heat conduction, functionally graded material, cellular materials

## INTRODUCTION

The FS-FEM was introduced in 2008 (Qin and Wang 2008; Wang and Qin 2009) and has become a very popular and powerful computational method in mechanical engineering. Unlike analytical solutions, which are available for only a few problems with simple geometries and boundary conditions (Qin and Mai 1997; Qin 1998; Qin and Mai 1998; Qin, Mai, and Yu 1999; Qin 2004; Qin and Ye 2004; Qin, Qu, and Ye 2005; Qu, Qin, and Kang 2006; Wang and Qin 2007; Yu and Qin 1996; Qin and Mai 1999), the FS-FEM is a highly efficient computational tool for the solution of various complex boundary value problems (Qin and Wang 2008; Wang and Qin 2017; Cao and Qin 2015; Wang and Qin 2019). So far, this method has been applied to potential problems (Wang, Qin, and Liang 2012; Gao, Wang, and Qin 2015; Wang and Qin 2010; Wang, Gao, and Qin 2015; Fu, Chen, and Qin 2011), plane elasticity (Wang and Qin 2012, 2011; Wang, Qin, and Yao 2012), wave propagation (Nanda 2020), composites (Qin and Wang 2015; Wang and Qin 2011; Wang and Qin 2011; Wang and Qin 2013; Cao, Qin, and Yu 2012; Wang and Qin 2015; Wang, Qin, and Xiao 2016), piezoelectric materials (Cao, Qin, and Yu 2012; Cao, Yu, and Qin 2013; Wang and Qin 2013), three dimensional problems (Cao, Qin, and Yu 2012; Lee, Wang, and Qin 2015), functionally graded materials (Cao, Wang, and Qin 2012; Cao, Qin, and Zhao 2012; Wang and Qin 2012; Wang, Qin, and Lee 2019), bioheat transfer problems (Wang and Qin 2010, 2012; Wang and Qin 2012; Zhang, Wang, and Qin 2012; Zhang, Wang, and Qin 2014; Tao, Qin, and Cao 2013; Zhang,

Wang, and Qin 2014), thermal elastic problems (Cao, Qin, and Yu 2012), hole problems (Qin and Wang 2013; Wang and Qin 2012), homogenization of heterogeneous elastic media (Meng, Gu, and Hager 2020), heat conduction problems (Qin and Wang 2011; Wang and Qin 2009), crack problems (Wang, Lin, and Qin 2019), topology optimization (Wang, Qin, and Lee 2019), axisymmetric potential problems (Zhou, Wang, and Li 2019), micromechanics problems (Cao, Qin, and Yu 2013; Cao, Yu, and Qin 2012), and anisotropic elastic problems (Cao, Yu, and Qin 2013; Cao, Yu, and Qin 2013; Wang and Qin 2010).

Following this introduction, basic formulations of FS-FEM and their applications to heat conduction problems are presented in the section "BASIC FORMULATION OF FS-FEM." Extensions to heat conduction problems in cellular solids with circular holes, functionally graded materials, and natural-hemp-fiber-filled cement composites are described in the subsequent 3 sections. Finally, a brief summary of the approach and future trends is provided.

## BASIC FORMULATION OF FS-FEM

Taking two-dimensional heat conduction as an example, this section discusses the fundamentals of FS-FEM. It is based on the work presented by Wang and Qin (2009). Beginning with a description of its basic equation, the construction of FS-FEM is discussed. This includes the derivation of the assumed intra-element field, auxiliary conforming frame field, modified variational principle, and recovery of rigid-body motion.

### Basic Equation of Heat Conduction

Consider a heat conduction problem in a general plane domain $\Omega$. Its field equation is written as (Wang and Qin 2009):

$$\frac{\partial}{\partial X_1}\left(k\frac{\partial u(z)}{\partial X_1}\right) + \frac{\partial}{\partial X_2}\left(k\frac{\partial u(z)}{\partial X_2}\right) = 0 \qquad \forall z(= x_1 + ix_2) \in \Omega \qquad (1)$$

and with the boundary conditions:
— Dirichlet boundary condition related to unknown temperature field

$$u = \bar{u} \qquad \text{on } \Gamma_u \qquad (2)$$

— Neumann boundary condition for the boundary heat flux

$$q = -ku_{,i}n_i = \bar{q} \qquad \text{on } \Gamma_q \qquad (3)$$

where $k$ stands for the thermal conductivity, $u$ is the sought field variable, and $q$ represents the boundary heat flux. $n_i$ is the $i$th component of the outward normal vector to the boundary $\Gamma = \Gamma_u \cup \Gamma_q$, and $\bar{u}$ $\bar{q}$ are specified functions on the related boundaries, respectively. Space derivatives are indicated by a comma, i.e., $u_{,i} = \partial u / \partial X_i$, and the subscript index $i$ takes values 1 and 2 in our analysis. The repeated subscript indices represent the summation convention. For convenience, Eq. (3) is rewritten in matrix form as:

$$q = -k\mathbf{A}\begin{bmatrix} u_{,1} \\ u_{,2} \end{bmatrix} = \bar{q} \qquad (4)$$

where $\mathbf{A} = \begin{bmatrix} n_1 & n_2 \end{bmatrix}$.

## Basic Formulation of FS-FEM

In this section, the procedure for developing a hybrid FEM with FS as an interior trial function is described, based on the boundary value problem defined by Eqs. (1)-(3).

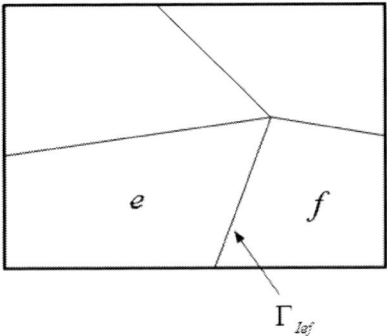

Figure 1. Illustration of continuity between two adjacent elements '$e$' and '$f$.'

As in the hybrid Trefftz FEM, the main idea of the present approach is to establish a hybrid FE formulation whereby intra-element continuity is enforced on a nonconforming internal displacement field formed by a linear combination of fundamental solutions at points outside the element domain under consideration, while an auxiliary frame field is independently defined on the element boundary to enforce field continuity across inter-element boundaries. But unlike in the hybrid Trefftz FEM, the intra-element field is constructed on the basis of the fundamental solution, rather than on T-functions. Consequently, a variational functional corresponding to the new trial function is required to derive the related stiffness matrix equation. With the problem domain divided into some subdomains or elements denoted by $\Omega_e$ with the element boundary $\Gamma_e$, additional continuity is usually required on the common boundary $\Gamma_{Ief}$ between any two adjacent elements '$e$' and '$f$' (see Figure 1):

$$\left.\begin{array}{l} u_e = u_f \quad \text{(conformity)} \\ q_e + q_f = 0 \quad \text{(reciprocity)} \end{array}\right\} \quad \text{on } \Gamma_{Ief} = \Gamma_e \cap \Gamma_f \qquad (5)$$

in the proposed hybrid FE approach.

## Nonconforming Intra-Element Field

Activated by the idea of the method of fundamental solution (MFS) (Wang and Qin 2008), to remove the singularity of the fundamental solution, for a particular element, say element $e$, which occupies the subdomain $\Omega_e$, we assume that the field variable defined in the element domain is extracted from a linear combination of fundamental solutions centered at different source points (see Figure 2), that is:

$$u_e(z) = \sum_{j=1}^{n_s} N_e(z, z_{0j}) c_{ej} = \mathbf{N}_e(z) \mathbf{c}_e \qquad \forall z \in \Omega_e, z_{0j} \notin \Omega_e \qquad (6)$$

where $c_{ej}$ is undetermined coefficients, $z_{0j} = x_{10} + ix_{20}$, and $n_s$ is the number of virtual sources outside the element $e$. $N_e(z, z_{0j})$ is the fundamental solution to the 2D heat conduction and generally satisfies

$$k\nabla^2 N_e(z, z_0) + \delta(z, z_0) = 0 \qquad \forall z, z_0 \in \square^2 \qquad (7)$$

which gives

$$N_e(z, z_0) = -\frac{1}{2\pi k} \operatorname{Re}\{\ln(z - z_0)\} \qquad (8)$$

where Re denotes the real part of the bracketed expression, $i = \sqrt{-1}$ the imaginary number. Clearly, Eq. (6) analytically satisfies Eq. (1) due to the solution property of $N_e(z, z_{0j})$.

In practice, the generation of virtual sources is usually done by means of the formulation employed in the MFS (Wang and Qin 2007, 2008):

$$z_0 = z_b + \gamma(z_b - z_c) \tag{9}$$

where $\gamma$ is a dimensionless coefficient, $z_b$ is the elementary boundary point and $z_c$ the geometrical centroid of the element. For the particular element shown in Figure 2, we can use the nodes of the element to generate related source points for simplicity.

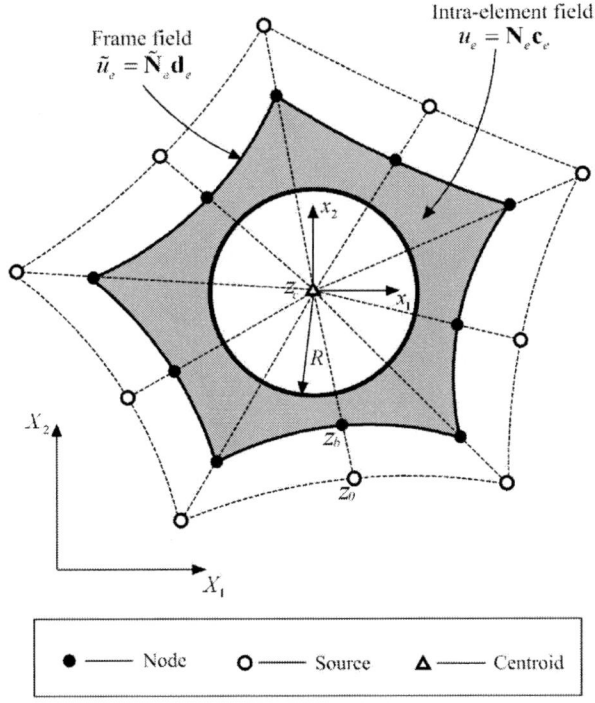

Figure 2. Intra-element field, frame field in a particular element in FS-FEM, and the generation of source points for a particular element.

The corresponding outward normal derivative of $u_e$ on $\Gamma_e$ is

$$q_e = -k\frac{\partial u_e}{\partial n} = \mathbf{Q}_e \mathbf{c}_e \tag{10}$$

where

$$\mathbf{Q}_e = -k\frac{\partial \mathbf{N}_e}{\partial n} = -k\mathbf{AT}_e \tag{11}$$

with

$$\mathbf{T}_e = \left\{\frac{\partial \mathbf{N}_e}{\partial x_1} \quad \frac{\partial \mathbf{N}_e}{\partial x_2}\right\}^T \tag{12}$$

## Auxiliary Conforming Frame Field

To enforce conformity on the field variable $u$, for instance, $u_e = u_f$ on $\Gamma_e \cap \Gamma_f$ of any two neighboring elements $e$ and $f$, an auxiliary inter-element frame field $\tilde{u}$ is used and expressed in terms of the same degrees of freedom (DOF), d, as used in the conventional finite element method. In this case, $\tilde{u}$ is confined to the entire element boundary, that is,

$$\tilde{u}_e(\mathbf{x}) = \tilde{\mathbf{N}}_e(z)\mathbf{d}_e \tag{13}$$

which is independently assumed along the element boundary in terms of nodal DOF $\mathbf{d}_e$, where $\tilde{\mathbf{N}}_e$ represents the conventional finite element interpolating functions. For example, a simple interpolation of the frame field on the side with three nodes of a particular element (Figure 2) can be given in the form:

$$\tilde{u} = \tilde{N}_1 u_1 + \tilde{N}_2 u_2 + \tilde{N}_3 u_3 \tag{14}$$

where $\tilde{N}_i$ ($i=1,2,3$) stands for shape functions in terms of the natural coordinate $\xi$ defined in Figure 3.

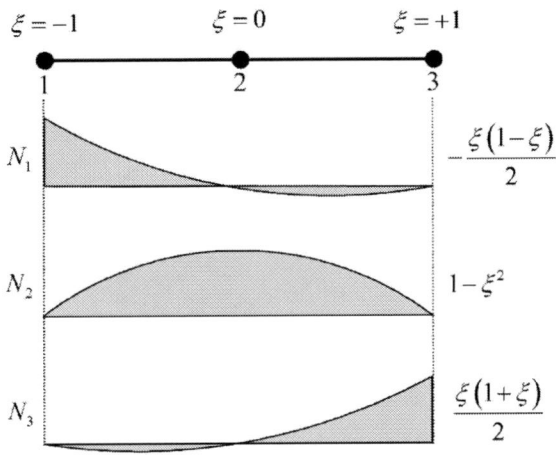

Figure 3. Typical quadratic interpolation for the frame field.

## Modified Variational Principle

For the boundary value problem defined in Eqs. (1)-(3) and (5), because the stationary conditions of the traditional potential or complementary variational functional cannot guarantee satisfaction of the inter-element continuity condition required in the proposed FS-FEM, a modified potential functional was developed by Qin and Wang (2008):

$$\Pi_m = \sum_e \Pi_{me} = \sum_e \left\{ \Pi_e + \int_{\Gamma_e} (\tilde{u} - u) q \, d\Gamma \right\} \tag{15}$$

where

$$\Pi_e = -\frac{1}{2}\int_{\Omega_e} ku_{,i}u_{,i}d\Omega - \int_{\Gamma_{qe}} \bar{q}\tilde{u}d\Gamma \qquad (16)$$

in which the governing Eq. (1) is assumed to be satisfied, a priori, for deriving the FS finite element model. The boundary $\Gamma_e$ of a particular element consists of the following parts:

$$\Gamma_e = \Gamma_{ue} \cup \Gamma_{qe} \cup \Gamma_{Ie} \qquad (17)$$

where $\Gamma_{Ie}$ represents the inter-element boundary of element 'e' shown in Figure 1.

## Stiffness Equation

Having independently defined the intra-element field and frame field in a particular element (see Figure 2), the next step is to generate the element stiffness equation through a variational approach.

Following the approach described by Qin and Wang (2008), the variational functional $\Pi_e$ corresponding to a particular element $e$ of the present problem can be written as

$$\Pi_{me} = -\frac{1}{2}\int_{\Omega_e} ku_{,i}u_{,i}d\Omega - \int_{\Gamma_{qe}} \bar{q}\tilde{u}d\Gamma + \int_{\Gamma_e} q(\tilde{u}-u)d\Gamma \qquad (18)$$

Appling the divergence theorem described by Qin and Wang (2008) to the above functional, we have the final functional for the FS finite element model:

$$\begin{aligned}\Pi_{me} &= \frac{1}{2}\left[\int_{\Gamma_e} qud\Gamma + \int_{\Omega_e} uk\nabla^2 ud\Omega\right] - \int_{\Gamma_{qe}} \bar{q}\tilde{u}d\Gamma + \int_{\Gamma_e} q(\tilde{u}-u)d\Gamma \\ &= -\frac{1}{2}\int_{\Gamma_e} qud\Gamma - \int_{\Gamma_{qe}} \bar{q}\tilde{u}d\Gamma + \int_{\Gamma_e} q\tilde{u}d\Gamma\end{aligned} \qquad (19)$$

Then, substitution of Eqs. (6), (10), and (13) into the functional (19) finally produces

$$\Pi_e = -\frac{1}{2}\mathbf{c}_e^T \mathbf{H}_e \mathbf{c}_e - \mathbf{d}_e^T \mathbf{g}_e + \mathbf{c}_e^T \mathbf{G}_e \mathbf{d}_e \tag{20}$$

in which

$$\mathbf{H}_e = \int_{\Gamma_e} \mathbf{Q}_e^T \mathbf{N}_e d\Gamma = \int_{\Gamma_e} \mathbf{N}_e^T \mathbf{Q}_e d\Gamma$$

$$\mathbf{G}_e = \int_{\Gamma_e} \mathbf{Q}_e^T \tilde{\mathbf{N}}_e d\Gamma, \quad \mathbf{g}_e = \int_{\Gamma_{qe}} \tilde{\mathbf{N}}_e^T \overline{q} d\Gamma$$

The symmetry of $\mathbf{H}_e$ is obvious from the definition (20) of the variational functional $\Pi_e$.

To enforce inter-element continuity on the common element boundary, the unknown vector $\mathbf{c}_e$ must be expressed in terms of nodal DOF $\mathbf{d}_e$. The minimization of the functional $\Pi_e$ with respect to $\mathbf{c}_e$ and $\mathbf{d}_e$ respectively yields

$$\frac{\partial \Pi_e}{\partial \mathbf{c}_e^T} = -\mathbf{H}_e \mathbf{c}_e + \mathbf{G}_e \mathbf{d}_e = 0$$
$$\frac{\partial \Pi_e}{\partial \mathbf{d}_e^T} = \mathbf{G}_e^T \mathbf{c}_e - \mathbf{g}_e = 0 \tag{21}$$

from which the optional relationship between $\mathbf{c}_e$ and $\mathbf{d}_e$ and the stiffness equation can be produced:

$$\mathbf{c}_e = \mathbf{H}_e^{-1} \mathbf{G}_e \mathbf{d}_e \quad \text{and} \quad \mathbf{K}_e \mathbf{d}_e = \mathbf{g}_e \tag{22}$$

where $\mathbf{K}_e = \mathbf{G}_e^T \mathbf{H}_e^{-1} \mathbf{G}_e$ stands for the element stiffness matrix.

It is worth pointing out that the evaluation of the right-handed vector $\mathbf{g}_e$ in Eq. (22) is the same as that in the conventional FEM, which is obviously convenient for implementation of the FS-FEM into an existing FEM program.

## Recovery of Rigid-Body Motion

Considering the physical definition of the fundamental solution, it is necessary to recover the missing rigid-body motion modes from the above results.

Following the method presented by Qin and Wang (2008), the missing rigid-body motion can be recovered by writing the internal potential field of a particular element $e$ as:

$$u_e = \mathbf{N}_e \mathbf{c}_e + c_0 \tag{23}$$

where the undetermined rigid-body motion parameter $c_0$ can be calculated using the least square matching of $u_e$ and $\tilde{u}_e$ at element nodes (Qin 2000):

$$\sum_{i=1}^{n} \left( \mathbf{N}_e \mathbf{c}_e + c_0 - \tilde{u}_e \right)^2 \Big|_{\text{node } i} = \min \tag{24}$$

which finally gives:

$$c_0 = \frac{1}{n} \sum_{i=1}^{n} \Delta u_{ei} \tag{25}$$

in which $\Delta u_{ei} = \left( \tilde{u}_e - \mathbf{N}_e \mathbf{c}_e \right)\big|_{\text{node } i}$ and $n$ is the number of element nodes.

Once the nodal field is determined by solving the final stiffness equation, the coefficient vector $\mathbf{c}_e$ can be evaluated from Eq. (22), and then $c_0$ is evaluated from Eq. (25). Finally, the potential field $u$ at any internal point in an element can be obtained by means of Eq. (6).

## FS-FEM FOR CELLULAR SOLIDS WITH CIRCULAR HOLES

As an application of the formulation presented, a special fundamental solution and the associated hole element are presented for modeling numerically cellular solids with circular holes in this section. All formulations in this section are taken from the work of Qin and Wang (2013).

### Fundamental Solutions

In order to perform numerical analysis accurately using the FS-FEM discussed here and to construct proper approximation fields, the fundamental solutions satisfying specified circular hole boundary conditions should be introduced.

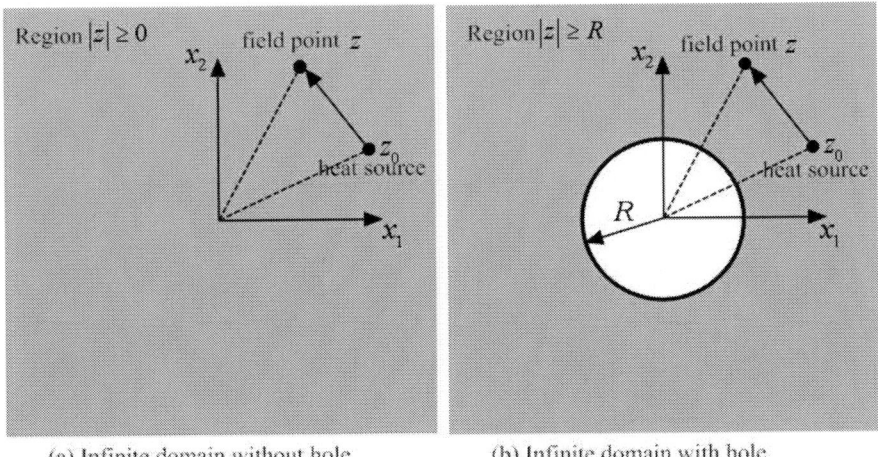

(a) Infinite domain without hole      (b) Infinite domain with hole

Figure 4. Demonstration of the definition of fundamental solutions.

## Fundamental Solution without Circular Hole

Consider a unit heat source located at the point $z_0$ in the infinite domain (Figure 4a). The fundamental solution $N$ is required to satisfy the governing equation (7) associated with a unit internal point source applied at the point $z_0$ in the complex space. Thus, the temperature response at any field point $z$ is given in the form of Eq. (8).

Obviously, the expression (8) shows singularity at $z = z_0$, which is an inherent feature of the fundamental solution.

## Fundamental Solution with a Centered Circular Hole

It is known that the special fundamental solution or Green's function refers to the singular solution that is required to satisfy not only the governing equation (7) but also specified boundary conditions. Here, we consider a unit heat source located at the point $z_0$ in the infinite domain containing a centered circular hole with radius $R$ (Figure 4b). In this case, the temperature response at any field point $z$ is given in the form (Ang 2007):

$$N(z, z_0) = \frac{1}{2\pi k} \text{Re}\left\{\ln\left(\frac{z-z_0}{R}\right)\right\} - \frac{1}{2\pi k} \text{Re}\left\{\ln\left(1 - \frac{z\bar{z}_0}{R^2}\right)\right\}$$
$$= \frac{1}{2\pi k} \text{Re}\left\{\ln\left(\frac{R(z-z_0)}{R^2 - z\bar{z}_0}\right)\right\} \quad (26)$$

for the case of $N = 0$ on the circular boundary, and

$$N(z, z_0) = \frac{1}{2\pi k} \text{Re}\left\{\ln\left(\frac{z-z_0}{R}\right)\right\} + \frac{1}{2\pi k} \text{Re}\left\{\ln\left(\frac{R^2 - z\bar{z}_0}{Rz}\right)\right\}$$
$$= \frac{1}{2\pi k} \text{Re}\left\{\ln\left(\frac{(z-z_0)(R^2 - z\bar{z}_0)}{R^2 z}\right)\right\} \quad (27)$$

for the case of $\partial N / \partial n = 0$ on the circular boundary.

## FS-FEM with Special Fundamental Solutions

In this subsection, the construction of special elements of the FS-FEM with the fundamental solutions (26) and (27) as an interior trial function is described, based on the boundary value problem defined by Eqs. (1)-(3). In the case of cellular materials with circular holes, the corresponding nonconforming intra-element field, auxiliary conforming frame field, and modified variational principle are still defined by Eqs. (6), (13), and (15), except that the special fundamental solution (26) or (27) is used to define the interpolation function $N_e(z, z_{0j})$ in Eq. (6).

In should be mentioned that one of the advantages in the presented FS-FEM is the reduction of computation time by using the special fundamental solution (26) or (27). Making use of two groups of independent interpolation functions in the FS-FEM, we can construct arbitrarily shaped elements for various engineering analyses. For convenience, we restrict our analysis to the following four element types (see Figure 5):

1. General 8-node quadrilateral element called E1, used for regions without any holes
2. Special purpose 8-node quadrilateral circular hole element called E2, used for regions with holes
3. Special purpose 16-node quadrilateral circular hole element called E3, used for regions with holes
4. Special purpose 16-node octagon circular hole element called E4, used for regions with holes.

Additionally, due to the geometrical symmetry of the circular hole, the specially constructed circular hole elements E2, E3, and E4 are also symmetrical; the side length is measured by a parameter $a$ and the diameter of the circular boundary is denoted as $D=2R$.

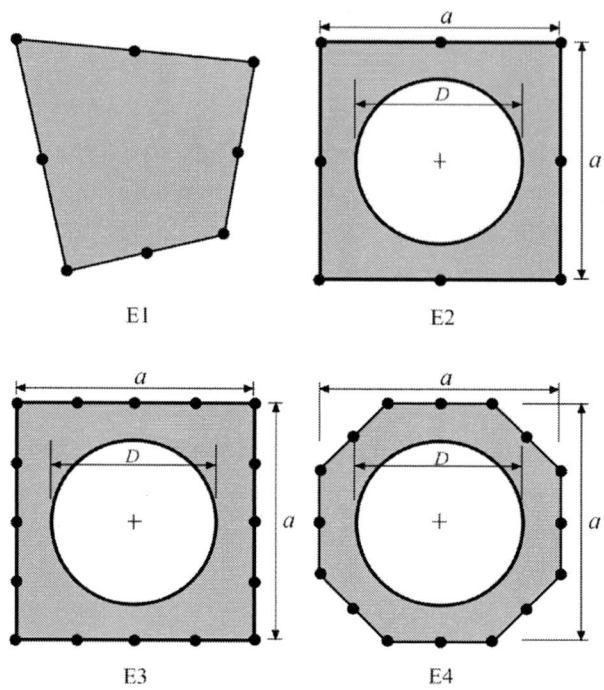

Figure 5. Types of elements constructed in FS-FEM.

## FS-FEM FOR GRADED MATERIALS

In this section we present a brief review of the results given by Cao, Wang, and Qin (2012). In doing this, consider a 2D heat conduction problem defined in an anisotropic inhomogeneous medium. In this case, Eq. (1) becomes

$$\frac{\partial \tilde{K}_{ij}}{\partial X_i}\frac{\partial u(z)}{\partial X_j}+\tilde{K}_{ij}\frac{\partial^2 u(z)}{\partial X_i \partial X_j}=0 \qquad \forall z \in \Omega \tag{28}$$

with the following boundary conditions:
— Specified temperature boundary condition

$$u = \bar{u} \quad \text{on } \Gamma_u \quad (29)$$

— Specified heat flux boundary condition

$$q = -\tilde{K}_{ij} u_{,j} n_j = \bar{q} \quad \text{on } \Gamma_q \quad (30)$$

where $\tilde{K}_{ij}$ denotes the thermal conductivity in terms of the spatial variable **X** and is assumed symmetric and positive-definite ($\tilde{K}_{12} = \tilde{K}_{21}, \det \tilde{K} = \tilde{K}_{11}\tilde{K}_{22} - \tilde{K}_{12}^2 > 0$). $u$ is the sought field variable and $q$ represents the boundary heat flux. $n_j$ is the direction cosine of the unit outward normal vector **n** to the boundary $\Gamma = \Gamma_u \cup \Gamma_q$, and $\bar{u}$ and $\bar{q}$ are specified functions on the related boundaries, respectively. For convenience, the space derivatives are indicated by a comma, i.e., $u_{,j} = \partial u / \partial X_j$, and the subscript index $i, j$ takes values 1 and 2 in our analysis. Moreover, the repeated subscript indices stand for the summation convention.

## Fundamental Solution in Functionally Graded Materials

For simplicity, we assume that the thermal conductivity varies exponentially with position vector, for example,

$$\tilde{\mathbf{K}}(\mathbf{X}) = \mathbf{K} \exp(2\boldsymbol{\beta} \cdot \mathbf{X}) \quad (31)$$

where vector $\boldsymbol{\beta} = (\beta_1, \beta_2)$ is a graded parameter and matrix K is symmetric and positive-definite with constant entries.

Substituting Eq. (31) into Eq. (28) yields

$$K_{ij} \partial_i \partial_j u(\mathbf{X}) + 2\beta_i K_{ij} \partial_j u(\mathbf{X}) = 0 \quad (32)$$

whose fundamental function defined in the infinite domain necessarily satisfies the equation

$$K_{ij}\partial_i\partial_j N(\mathbf{X},\mathbf{X}_s) + 2\beta_i K_{ij}\partial_j N(\mathbf{X},\mathbf{X}_s) + \delta(\mathbf{X},\mathbf{X}_s) = 0 \tag{33}$$

in which X and $X_s$ denote an arbitrary field point and a source point in the infinite domain, respectively. $\delta$ is the Dirac delta function.

The closed-form solution to Eq. (33) in two dimensions can be expressed as (Berger et al. 2005):

$$N(\mathbf{X},\mathbf{X}_s) = -\frac{K_0(\kappa R)}{2\pi\sqrt{\det \mathbf{K}}} \exp\{-\boldsymbol{\beta}\cdot(\mathbf{X}+\mathbf{X}_s)\} \tag{34}$$

where $\kappa = \sqrt{\boldsymbol{\beta}\cdot\mathbf{K}\boldsymbol{\beta}}$, $R$ is the geodesic distance defined as $R = R(\mathbf{X},\mathbf{X}_s) = \sqrt{\mathbf{r}\cdot\mathbf{K}^{-1}\mathbf{r}}$ and $\mathbf{r} = \mathbf{r}(\mathbf{X},\mathbf{X}_s) = \mathbf{X}-\mathbf{X}_s$. $K_0$ is the modified Bessel function of the second kind of zero order. For isotropic materials, $K_{12} = K_{21} = 0$, $K_{11} = K_{22} = k_0 > 0$, Eq. (32) is recast as

$$k_0 \nabla^2 u(\mathbf{X}) + 2k_0 \beta_i \partial_i u(\mathbf{X}) = 0 \tag{35}$$

Then the fundamental solution given by Eq. (34) reduces to

$$N(\mathbf{X},\mathbf{X}_s) = -\frac{K_0(\kappa R)}{2\pi k_0} \exp\{-\boldsymbol{\beta}\cdot(\mathbf{X}+\mathbf{X}_s)\} \tag{36}$$

which agrees with the result in Gray et al. (2003).

## Generation of Graded Element

Having obtained the fundamental solutions for graded materials, the procedure for developing the corresponding hybrid graded element model is similar to that described in "Basic formulation of FS-FEM." In the case of graded materials, the corresponding nonconforming intra-element field, auxiliary conforming frame field, and modified variational principle are still defined by Eqs. (6), (13), and (15), respectively, except that the special fundamental solution (34) or (36) is used to define the interpolation function $N_e(z, z_{0j})$ in Eq. (6).

In the following discussion, the fundamental solution in Eq. (34) for functionally graded materials (FGMs) is used as $N_e$ of Eq. (6) to approximate the intra-element field. It can be seen from Eq. (34) that $N_e$ varies throughout each element due to different geodesic distances for each field point, so smooth variation of material properties can be achieved by this inherent property, instead of by the stepwise constant approximation frequently used in the conventional FEM. For example, Figure 6 illustrates two models in which the thermal conductivity varies along the direction $X_2$ in isotropic material.

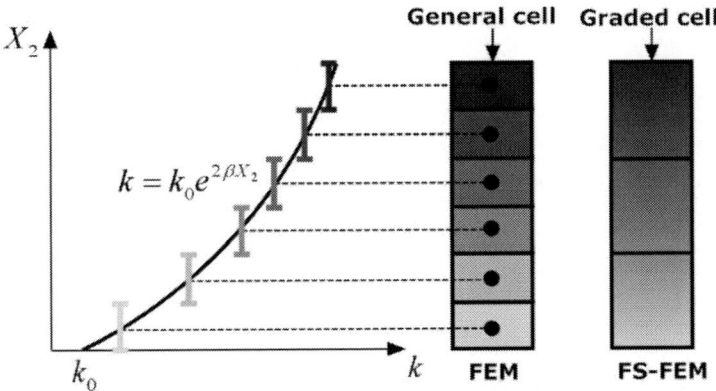

Figure 6. Comparison of computational cell in the conventional FEM and the proposed FS-FEM.

It should be mentioned here that Eq. (31), which describes variation of thermal conductivity, is defined under a global coordinate system. When contriving the intra-element field for each element, this formulation must be transferred into a local element coordinate defined at the center of the element. The graded matrix $\tilde{\mathbf{K}}$ in Eq. (31) can then be expressed by

$$\tilde{\mathbf{K}}_e(\mathbf{x}) = \mathbf{K}_C \exp(2\boldsymbol{\beta} \cdot \mathbf{x}) \tag{37}$$

for a particular element $e$, where $K_c$ denotes the value of conductivity at the centroid of each element and can be calculated as follows:

$$\mathbf{K}_C = \mathbf{K} \exp(2\boldsymbol{\beta} \cdot \mathbf{X}_c) \tag{38}$$

where $X_c$ is the global coordinate of the element centroid.

Accordingly, the matrix $K_c$ is used to replace K (see Eq. (34) in the formulation of fundamental solution for FGMs and the construction of intra-element field under local element coordinate for each element.

## SPECIAL N-NODED VORONOIFIBER/MATRIX ELEMENT

In this section, the developments in Wang, Qin, and Xiao 2016) are used for analyzing thermal effects of clustering in natural-hemp-fiber filled cement composites and determining effective thermal conductivity of the composites. Beginning with a discussion of a micromechanical model of clustered composites, a special n-noded Voronoi fiber/matrix element is constructed. Numerical examples are presented to demonstrate the efficiency of these special elements in dealing with clustering distribution of fibers.

### Micromechanical Model of Clustered Composite

In a periodic cement-based composite containing clustered hemp fibers, the representative unit cell is the smallest repeated microstructure of the composite that can be isolated from the composite to estimate the composite's effective properties. It is assumed that the unit cell has same thermal properties and fiber volume contents as the composite under consideration.

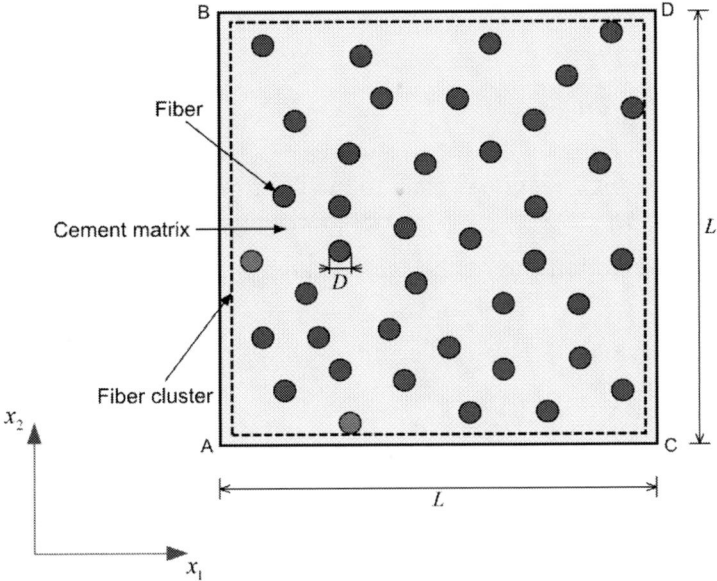

Figure 7. Schematic representation of regularly and randomly clustered fibers within the cement matrix.

Figure 7 shows a representative unit cell containing clustered hemp fibers. In Figure 7, $L$ denotes the cell length, $D$ is the diameter of the hemp fiber, and $x_1$ and $x_2$ are the global coordinate axial directions. Under the assumptions that (1) all material constituents are isotropic and homogeneous, and (2) the hemp fiber and the cement matrix are perfectly bonded, the steady-state local temperature fields in the matrix and the fiber, denoted by $T_m$ and $T_f$, should satisfy the two-dimensional heat conduction governing equations respectively, given by

$$\frac{\partial^2 T_m}{\partial x_1^2} + \frac{\partial^2 T_m}{\partial x_2^2} = 0, \quad \frac{\partial^2 T_f}{\partial x_1^2} + \frac{\partial^2 T_f}{\partial x_2^2} = 0 \tag{39}$$

and the continuous conditions at the interface between the hemp fiber and the matrix

$$T_m = T_f$$
$$k_m \frac{\partial T_m}{\partial n} = k_f \frac{\partial T_f}{\partial n} \qquad (40)$$

where $n$ is the unit direction normal to the fiber/matrix interface.

According to Fourier's law of heat transfer in isotropic media, we have the following relationship of the temperature variable $T$ and the heat flux component $q_i$:

$$q_i = -k \frac{\partial T}{\partial x_i} \qquad (i = 1, 2) \qquad (41)$$

from which the effective thermal conductivity $k_e$ of the homogenized composite can be determined by

$$k_e = \frac{\bar{q}_i}{\bar{\varepsilon}_i} \qquad (42)$$

where $\bar{q}_i$ stands for the area-averaged heat flux component along the $i$-direction and $\bar{\varepsilon}_i$ the temperature gradient component along the $i$-direction. For example, for the applied temperature boundary conditions:

$$\begin{aligned} T_m &= T_0 & &\text{on edge AB} \\ T_m &= T_1 & &\text{on edge CD} \\ k_m \frac{\partial T_m}{\partial n} &= 0 & &\text{on edges AC and BD} \end{aligned} \qquad (43)$$

The effective thermal conductivity $k_e$ of the composite can be calculated by the 1-directional average heat flux component $\bar{q}_1$ on the surface CD and the 1-directional temperature gradient component $\bar{\varepsilon}_1$ respectively given by

$$\bar{q}_1 = \frac{1}{L}\int_{AB} q_1(x_1, x_2) dx_2 \tag{44}$$

$$\bar{\varepsilon}_1 = \frac{(T_1 - T_2)}{L} \tag{45}$$

## Special $n$-Sided Voronoi Fiber/Matrix Element

The representative unit cell with the specified temperature conditions along the outer boundary of the cell is solved by a fundamental solution based hybrid finite element formulation with special $n$-sided Voronoi fiber/matrix elements. To efficiently treat regularly and randomly clustered distributions of hemp fibers in the unit cell and obtain a mesh with relatively high quality, the centroidal Voronoi tessellation technique is employed such that the generators for the Voronoi tessellation and the centroids of the Voronoi regions coincide (Du, Faber, and Gunzburger 1999). The centroidal Voronoi tessellation technique can be viewed as an optimal partition corresponding to an optimal distribution of generators. Figure 8 displays a typical $n$-sided Voronoi fiber/matrix element division for the composite cell including hemp fiber and cement material constituents. As an example, in Figure 2, the centroidal Voronoi elements are iteratively generated by the MATLAB source code (Wang, Qin, and Xiao 2016) using 25 random points in the cell and the fibers are located at the centroids of the Voronoi elements.

In a typical $n$-sided Voronoi fiber/matrix element $e$, with element domain $\Omega_e$ and element boundary $\Gamma_e$, the assumed fields include:

(a) Nonconforming interior temperature field

$$T(\mathbf{x}) = \sum_{j=1}^{m} G(\mathbf{x}, \mathbf{x}_{sj}) c_{ej} = \mathbf{N}_e \mathbf{c}_e \qquad \mathbf{x} \in \Omega_e \tag{46}$$

(b) Auxiliary conforming frame temperature field

$$\tilde{T}(\mathbf{x}) = \tilde{\mathbf{N}}_e \mathbf{d}_e \qquad \mathbf{x} \in \Gamma_e \qquad (47)$$

where $G$ is the fundamental solutions satisfying equilibrium and continuity within the element, $\mathbf{x}(x_1, x_2)$ and $\mathbf{x}_{sj}(x_{1j}^s, x_{2j}^s)$ are the field point and source point, respectively, $\mathbf{N}_e$ is a row vector of fundamental solutions, $\mathbf{c}_e$ is a column vector of the unknown coefficient $c_{ej}$, $\tilde{\mathbf{N}}_e$ represents a row vector of the conventional interpolating shape functions, and $\mathbf{d}_e$ is a column vector of the nodal degree of freedom of the element.

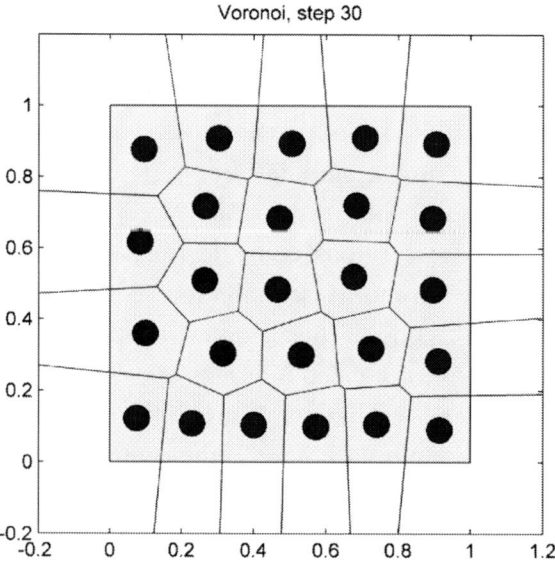

Figure 8. Schematic diagram of *n*-sided Voronoi fiber/matrix elements for the case of a random cluster.

Subsequently, the heat flux field in the element can be derived by means of Fourier's law

$$q_i(\mathbf{x}) = -k \frac{\partial T(\mathbf{x})}{\partial x_i} = \mathbf{T}_{ei} \mathbf{c}_e \qquad \mathbf{x} \in \Omega_e \qquad (48)$$

with

$$\mathbf{T}_{ei} = -k \frac{\partial \mathbf{N}_e}{\partial x_i} \qquad (49)$$

Furthermore, the outward normal heat flux $q_n$ derived from the interior field $T_e$ can be expressed as

$$q_n = q_1 n_1 + q_2 n_2 = \mathbf{Q}_e \mathbf{c}_e \qquad (50)$$

with

$$\mathbf{Q}_e = \mathbf{T}_{e1} n_1 + \mathbf{T}_{e2} n_2 \qquad (51)$$

and $n_i$ ($i=1,2$) are components of the outward unit normal to the element boundary.

To link the two independent fields described, the element variational functional (15) is now of the form (Wang and Qin 2009; Qin and Wang 2008)

$$\Pi_{me} = -\frac{1}{2} \int_{\Omega_e} k \left( \frac{\partial^2 T}{\partial x_1^2} + \frac{\partial^2 T}{\partial x_2^2} \right) d\Omega - \int_{\Gamma_{eq}} \bar{q} \tilde{T} d\Gamma + \int_{\Gamma_e} q_n \left( \tilde{T} - T \right) d\Gamma \qquad (52)$$

in which $\Gamma_{eq}$ is the element heat flux boundary, and $\bar{q}$ is the specified normal heat flux.

Application of the divergence theorem to the functional (52) yields

$$\Pi_{me} = -\frac{1}{2} \int_{\Gamma_e} q_n T d\Gamma - \int_{\Gamma_{qe}} \bar{q} \tilde{T} d\Gamma + \int_{\Gamma_e} q_n \tilde{T} d\Gamma \qquad (53)$$

Then, substituting Eqs. (46) and (47) into the functional (53) yields Eq. (20). The minimization of Eq. (20) leads to the element stiffness equation (22).

Clearly, evaluation of $K_e$ in Eq. (22) involves inversion of the symmetric square $H_e$ matrix, and it is advantageous to use a minimum number of source points outside the element for improving inversion efficiency. In contrast, accuracy is generally improved if a large number of source points is used. In this study, the number of source points is chosen to be same as that of element nodes, to balance the requirements of efficiency and accuracy. Certainly, the rank sufficiency condition in the hybrid finite element method (Ghosh 2011; Qin 2000, 1995; Pian and Wu 2005) is also satisfied. Moreover, unlike the Voronoi cell finite element method (Ghosh 2011), the present method is a type of hybrid displacement finite element method and all integrals involved are along the element boundary only. However, the present method requires fundamental solutions of the related problem, which are not available for some physical problems.

The following two-component heterogeneous fundamental solutions satisfying the equilibrium and continuity of fiber and matrix domains can be written as (Wang and Qin 2011)

$$G(z,z_{sj}) = -\frac{1}{2\pi k_m}\left\{\text{Re}\left[\ln(z-z_{sj})\right] + \frac{k_m - k_f}{k_m + k_f}\text{Re}\left[\ln(\frac{R^2}{z} - \bar{z}_{sj})\right]\right\} \quad \text{for } z \in \Omega_m$$

$$G(z,z_{sj}) = -\frac{1}{(k_m + k_f)\pi}\text{Re}\left[\ln(z-z_{sj})\right] \quad \text{for } z \in \Omega_f$$

(54)

where $z = x_1 + ix_2$ and $z_{sj} = x_{1j}^s + ix_{2j}^s$ are the complex coordinates of the field point and the source point, respectively, $R$ is the radius of fiber inclusion, and $i = \sqrt{-1}$ is the imaginary unit. $\Omega_m$ and $\Omega_f$ are respectively the cement matrix domain and the hemp fiber domain. In particular, if $k_m = k_f = k$, Eq. (54) reduces to

$$G(z,z_{sj}) = -\frac{1}{2k\pi}\text{Re}\left[\ln(z-z_{sj})\right] \quad (55)$$

which corresponds to the heat transfer caused by a point heat source in an isotropic homogeneous medium.

## Results and Discussion

Due to the clustered distribution of the hemp fibers, the global fiber volume fraction is defined by

$$v_{fc} = \frac{p\pi R^2}{L \times L} \tag{56}$$

where $p$ is the number of hemp fibers embedded in the unit cell and $R$ is the radius of the hemp fiber. In practical computation, the radius of the hemp fiber is taken as $R = 100\mu m$ (Liu, Takagi, and Yang 2011), and the number of fibers is assumed to be $p=25$. It should be understood that although only 25 fibers are involved in the computational model, our approach can be used to solve more complex problems involving large numbers of fibers, i.e., thousands of fibers, with incorporation of the centroidal Voronoi diagram. The thermal conductivities of the fiber and the cement material are $k_f = 0.115$ W/mK (Behzad and Sain 2007) and $k_m = 0.53$ W/mK (Xu and Chung 2000), respectively.

### *Validation of Element Property*

To rigorously investigate and understand the effect of clustered fibers on the effective thermal conductivity of the composite, the results of well-dispersed fiber distribution involving single hemp fibers are provided as reference values for comparison. To demonstrate the convergence and accuracy of the present special $n$-sided fiber/matrix element, the unit cell shown in Figure 9 is respectively modeled using a special 4-sided fiber/matrix element, a special 8-sided fiber/matrix element, and a special 12-sided fiber/matrix element. Correspondingly, the number of nodes is 8, 16, and 24, respectively. This means that there are 3 nodes along each element side and the quadratic shape function interpolation is employed for approximating the conforming frame temperature field along the element boundary. Moreover, because only one special element is used for each simulation, the global fiber volume fraction to the cell here is also the local fiber volume fraction to the element.

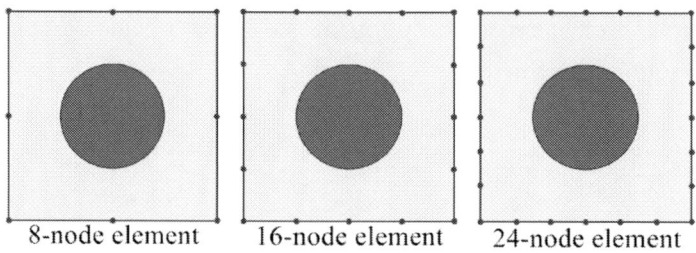

Figure 9. Schematic diagram of special elements for the cement composite unit cell with well-dispersed hemp fibers.

The numerical model is first validated by assuming $k_m = k_f = k = 0.53$ W/(mK). As expected, the predicted effective thermal conductivity of the composite falls within the errors of 0.0057%, 0.0019%, and 0.00006% to the theoretical value given by $k^{\text{theoretical}} = 0.53$ W/(mK) for the present three $n$-sided special fiber/matrix element types, and all results are independent of fiber volume fractions. It is found that the difference between the numerical and theoretical results decreases along with the increase in the number of element sides or nodes.

Next, Figure 10 displays the percentage difference in the effective thermal conductivity between the results of the present $n$-sided special fiber/matrix elements and the converged finite element results for the fiber volume fraction ranging from 0.05 to 0.72. Note that the converged finite element results are obtained by ABAQUS with extremely refined 8-node finite elements (DC2D8) ranging from 3036 to 16797, depending on the dimensions of the core fiber and the cell. The core fiber and the matrix thermal conductivities are respectively taken as $k_f = 0.115$ W/(mK) and $k_m = 0.53$ W/(mK). Figure 10 indicates that the present special $n$-sided fiber/matrix element has good convergence when the number of sides or nodes of the special element increases. Also, it is seen that the special 12-sided fiber/matrix element can give the best accuracy for a large range of the fiber volume fraction and the percent difference is only 1.75% when the fiber volume fraction is equal to 0.72, which is very close to the value when the fiber is tangential to the cell boundary.

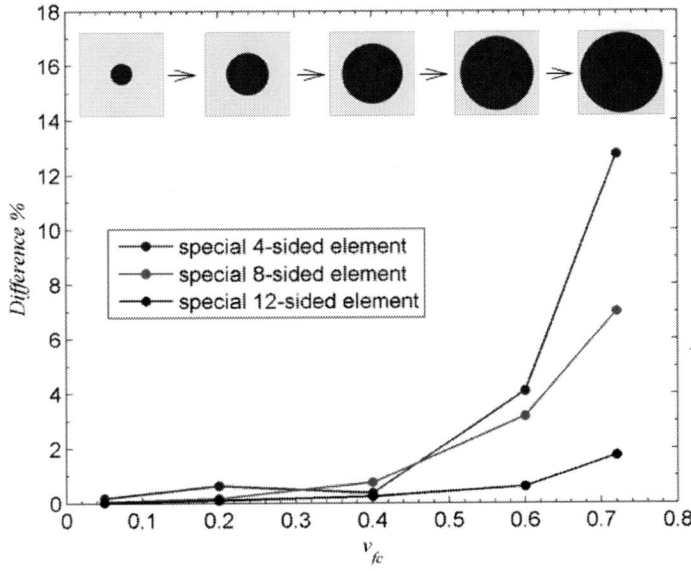

Figure 10. Convergence of the present special $n$-sided fiber/matrix elements.

## *Effect of Degree of Fiber Clustering*

To perform a parametric study of the clustering effect of the hemp fiber embedded in the cement matrix, the centroid Voronoi tessellation technique is used to obtain the $n$-sided area polygons encompassing a given set of generators, which denotes the set of fiber centers for the present work. Generally, the number of identical hemp fibers embedded in the cement matrix can be arbitrary. Here, 25 identical hemp fibers are arranged within the matrix in the manner shown in Figure 11 to represent clustering. In Figure 11, four clustered cases, including three regular cases and one random case, are considered for the purpose of comparison. The global fiber volume content is taken to be $v_{fc} = 10\%$, so that the degree of fiber clustering can be adjusted over a relatively large range. The thermal conductivities of the hemp fiber and the cement matrix are taken as $k_f = 0.115$ W/mK and $k_m = 0.53$ W/mK, respectively.

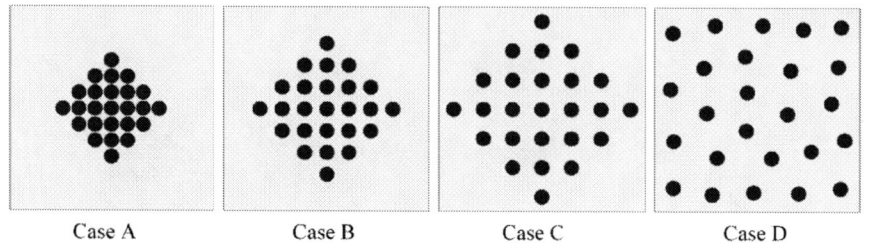

Figure 11. Distributions for four clustered cases at 10% fiber global volume content using 25 hemp fibers.

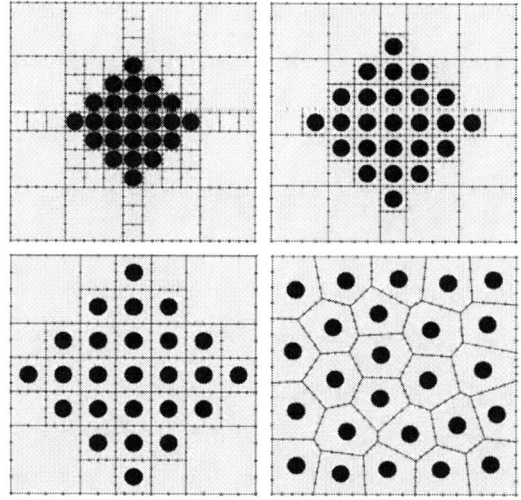

Figure 12. Mesh configurations for four clustering cases using special $n$-sided elements.

Figure 12 displays mesh divisions with the present $n$-sided polygonal elements for modeling the unit composite cell. For example, for the case D representing the randomly clustered arrangement of fibers, five types of Voronoi polygonal elements with 16 to 24 nodes are involved. It is necessary to point out that some artificial points have been added to produce regular Voronoi diagrams for the cases A, B, and C, and for those elements without a fiber phase the setting $k_m = k_f = k = 0.53$ W/(mK) is required. To clearly demonstrate the effect of the degree of clustering on the effective thermal conductivity $k_e$ of the composite, the percent difference of results from the four cluster cases and the well-dispersed case without a cluster is shown in Figure 13 and the reference value is taken as 0.4659 W/mK for error analysis, which corresponds to the result of the well-dispersed case by FEM. It is observed that case A, which corresponds to the greatest degree of clustering, produces the largest deviation from the well-dispersed case, but the percent difference is less than 0.6%. Moreover, Figure 13 again shows good agreement between the results from FEM and the present $n$-sided elements.

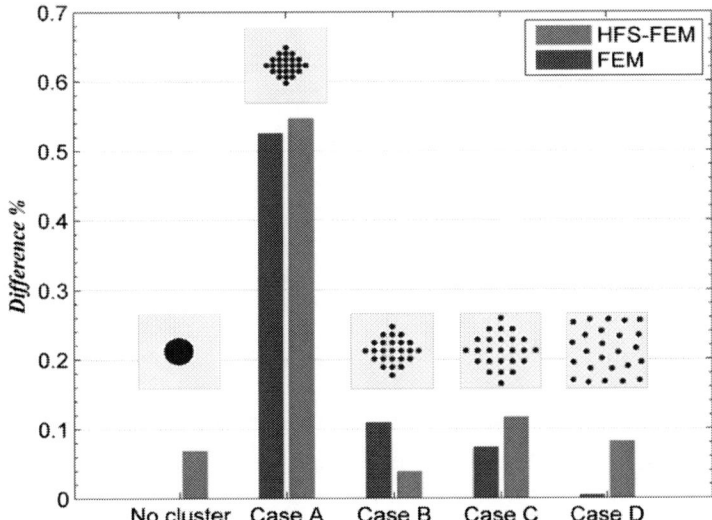

Figure 13. Effect of clustering on the effective thermal conductivity of the composite relative to the well-dispersed fiber distribution at 10% global fiber volume fraction.

## *Effect of Global Fiber Volume Fraction*

Figure 14 demonstrates the effect of clustering on the effective thermal conductivity $k_e$ of the composite for a range of global fiber volume fractions (5%-40%) for a random cluster arrangement, i.e., case D. Generally, the same decreasing effect can be observed for both clustered and well-dispersed fiber arrangements in the cement matrix with increasing global volume fraction. This is due to the fact that the thermal conductivity of the hemp fiber is obviously lower than that of the cement matrix. Further, it is found from Figure 14 that the difference between the clustered and the well-dispersed fiber arrangements is significantly augmented when the global fiber volume fraction is about 20% and the corresponding percent difference is 1.95%.

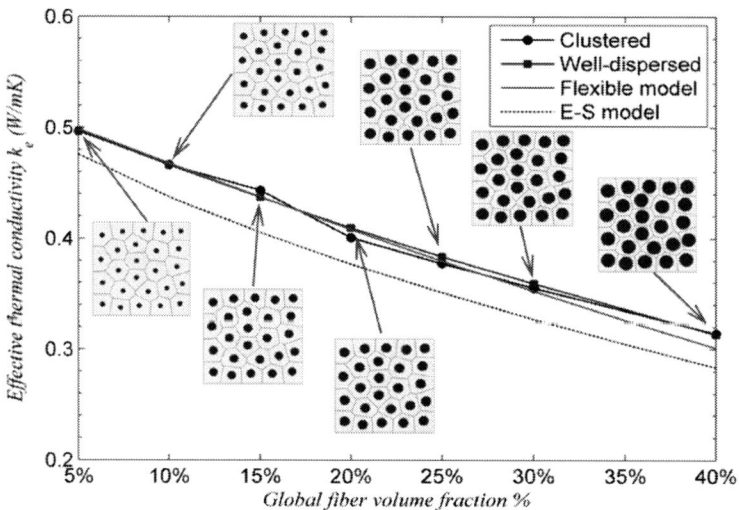

Figure 14. Effect of global fiber volume fraction for the randomly clustered case D.

In addition to the numerical results, Figure 14 also lists, respectively, the theoretical predictions from the flexible model (Kirkpatrick 1973)

$$k_e = \frac{1}{f-2}\left\{F^* + \sqrt{F^{*2} + 2(f-2)k_m k_f}\right\}, \tag{57}$$

where $F^* = (\frac{f}{2}v_m - 1)k_m + (\frac{f}{2}v_{fc} - 1)k_f$, the flexible factor $f = 4.5$, and the E-S model (Zou et al. 2003)

$$k_e = k_m \left[ 1 - \frac{1}{c} + \frac{\pi}{2d} - \frac{c}{d\sqrt{d^2 - c^2}} \ln\left( \frac{d + \sqrt{d^2 - c^2}}{c} \right) \right] \quad (58)$$

with $c = \sqrt{\pi / v_{fc}} / 2$, $d = 1/\beta - 1$ and $\beta = k_f / k_m$. It is clearly observed from Figure 14 that the flexible model can match the numerical results better than the E-S model within the range of fiber volume fraction shown in the figure.

## CONCLUSION AND FUTURE DEVELOPMENTS

A new type of fundamental solution based FEM formulation has been presented for analyzing two-dimensional heat conduction problems, with application to heat conduction problems in cellular solids with circular holes, functionally graded materials, and natural-hemp-fiber-filled cement composites. In the model, a linear combination of the fundamental solution at points outside the element domain is used to approximate the field variable within an element domain, and a frame field defined on the elementary boundary is introduced to guarantee inter-element continuity. To adapt the new trial function (fundamental solution) a modified variational functional is introduced to establish the corresponding stiffness matrix equation.

In contrast to conventional FE and boundary element models, the main advantages of the FS FE model are: (i) the formulation calls for integration along the element boundaries only, which enables arbitrary polygonal or even curve-sided elements to be generated. As a result, it may be considered a special, symmetric, substructure-oriented boundary solution approach, which thus possesses the advantages of the BEM. (ii) the FS FE model is likely to represent the optimal expansion bases for hybrid-type elements where inter-element continuity need not be satisfied, a priori, which is particularly important for generating a quasi-conforming plate-bending element; (iii) the model offers the attractive possibility of developing accurate crack singular, corner, or perforated elements, simply by using appropriate known local solution functions as the trial functions of the intra-element displacements.

It is recognized that the FS FE method has become increasingly popular as an efficient numerical tool in computational mechanics since its initiation in the late early 2000s. However, many possible extensions and areas are still in need of further development. Among those developments one could list the following:

1. Development of efficient FS FE-BEM schemes for complex engineering structures and the related general-purpose computer codes with pre-processing and postprocessing capabilities.
2. Generation of various special purpose functions to effectively handle singularities attributable to local geometrical or load effects. As discussed previously, the special purpose functions ensure that excellent results are obtained at minimal computational cost and without local mesh refinement. Extensions of such functions could be applied to other cases, such as the boundary layer effect between two materials, the interaction between fluid and structure in fluid–structure problems, and circular hole, corner, and load singularities.
3. Development of FS FE in conjunction with a topology optimization scheme and machine learning to contribute to microstructure and metamaterial design.
4. Development of efficient adaptive procedures including error estimation, h-extension elements, higher order p-capabilities, and convergence studies.
5. Extensions of FS FE to dynamic problems of soil mechanics, deep shell structure, fluid flow, piezoelectric materials, and rheology problems.

## REFERENCES

Ang, W. T. 2007. *A beginner's course in boundary element methods.* Boca Raton: Universal-Publishers.

Behzad, T., and M. Sain. 2007. Measurement and prediction of thermal conductivity for hemp fiber reinforced composites. *Polymer Engineering & Science* 47 (7):977-983.

Berger, J., P. Martin, V. Mantič, and L. Gray. 2005. Fundamental solutions for steady-state heat transfer in an exponentially graded anisotropic material. *Zeitschrift für angewandte Mathematik und Physik ZAMP* 56 (2):293-303.

Cao, C., and Q. H. Qin. 2015. Hybrid fundamental solution based finite element method: theory and applications. *Advances in Mathematical Physics* 2015:916029.

Cao, C., Q. H. Qin, and A. Yu. 2012. A new hybrid finite element approach for three-dimensional elastic problems. *Archives of Mechanics* 64 (3):261-292.

Cao, C., Q. H. Qin, and A. Yu. 2012. A novel boundary-integral based finite element method for 2D and 3D thermo-elasticity problems. *Journal of Thermal Stresses* 35 (10):849-876.

Cao, C., Q. H. Qin, and A. Yu. 2012. Hybrid fundamental-solution-based FEM for piezoelectric materials. *Computational Mechanics* 50 (4):397-412.

Cao, C., Q. H. Qin, and A. Yu. 2012. Modelling of Anisotropic Composites by Newly Developed HFS-FEM. *Paper read at Proceedings of the 23rd International Congress of Theoretical and Applied Mechanics*, Yilong Bai, Jianxiang Wang, Daining Fang (eds), SM08-016, at Beijing, China.

Cao, C., Q. H. Qin, and A. Yu. 2013. Micromechanical Analysis of Heterogeneous Composites using Hybrid Trefftz FEM and Hybrid Fundamental Solution Based FEM. *Journal of Mechanics* 29 (4):661-674.

Cao, C., A. Yu, and Q. H. Qin. 2013. A new hybrid finite element approach for plane piezoelectricity with defects. *Acta Mechanica* 224 (1):41-61.

Cao, C., A. Yu, and Q. H. Qin. 2013. A novel hybrid finite element model for modeling anisotropic composites. *Finite Elements in Analysis and Design* 64:36-47.

Cao, C., A. Yu, and Q. H. Qin. 2012. Evaluation of effective thermal conductivity of fiber-reinforced composites by boundary integral based finite element method. *International Journal of Architecture, Engineering and Construction* 1 (1):14-29.

Cao, C., A. Yu, and Q. H. Qin. 2013. Mesh reduction strategy: Special element for modelling anisotropic materials with defects. *Boundary Elements and Other Mesh Reduction Methods XXXVI* 56:61.

Cao, L. L., Q. H. Qin, and N. Zhao. 2012. Hybrid graded element model for transient heat conduction in functionally graded materials. *Acta Mechanica Sinica* 28 (1):128-139.

Cao, L., H. Wang, and Q. H. Qin. 2012. Fundamental solution based graded element model for steady-state heat transfer in FGM. *Acta Mechanica Solida Sinica* 25 (4):377-392.

Du, Q., V. Faber, and M. Gunzburger. 1999. Centroidal Voronoi tessellations: Applications and algorithms. *SIAM review* 41 (4):637-676.

Fu, Z. J., W. Chen, and Q. H. Qin. 2011. Hybrid Finite Element Method Based on Novel General Solutions for Helmholtz-Type Problems. *Computers Materials and Continua* 21 (3):187-208.

Gao, Y. T., H. Wang, and Q. H. Qin. 2015. Orthotropic Seepage Analysis using Hybrid Finite Element Method. *Journal of Advanced Mechanical Engineering* 2 (1):1-13.

Ghosh, S. 2011. *Micromechanical analysis and multi-scale modeling using the Voronoi cell finite element method*: CRC Press.

Gray, L., T. Kaplan, J. Richardson, and G. H. Paulino. 2003. Green's functions and boundary integral analysis for exponentially graded materials: heat conduction. *J. Appl. Mech.* 70 (4):543-549.

Kirkpatrick, S. 1973. Percolation and conduction. *Reviews of modern physics* 45 (4):574-588.

Lee, C. Y., H. Wang, and Q. H. Qin. 2015. Method of fundamental solutions for 3D elasticity with body forces by coupling compactly supported radial basis functions. *Engineering Analysis with Boundary Elements* 60:123-136.

Liu, K., H. Takagi, and Z. Yang. 2011. Evaluation of transverse thermal conductivity of Manila hemp fiber in solid region using theoretical method and finite element method. *Materials & Design* 32 (8-9):4586-4589.

Meng, C. F., C. Gu, and B. Hager. 2020. An Eshelby Solution-Based Finite-Element Approach to Heterogeneous Fault-Zone Modeling. *Seismological Research Letters* 91 (1):465-474.

Nanda, N. 2020. Spectral finite element method for wave propagation analysis in smart composite beams containing delamination. *Aircraft Engineering and Aerospace Technology* 92 (3):440-451.

Pian, T. H., and C. C. Wu. 2005. *Hybrid and incompatible finite element methods*: CRC press.

Qin, Q. H. 1995. Hybrid-Trefftz finite element method for Reissner plates on an elastic foundation. *Computer Methods in Applied Mechanics and Engineering* 122 (3-4):379-392.

Qin, Q. H. 1998. Thermoelectroelastic Green's function for a piezoelectric plate containing an elliptic hole. *Mechanics of Materials* 30 (1):21-29.

Qin, Q. H. 2000. *The Trefftz finite and boundary element method*: Southampton, WIT Press.

Qin, Q. H. 2004. Green's functions of magnetoelectroelastic solids with a half-plane boundary or bimaterial interface. *Philosophical Magazine Letters* 84 (12):771-779.

Qin, Q. H., and Y. W. Mai. 1997. Crack growth prediction of an inclined crack in a half-plane thermopiezoelectric solid. *Theoretical and Applied Fracture Mechanics* 26 (3):185-191.

Qin, Q. H., and Y. W. Mai. 1998. Thermoelectroelastic Green's function and its application for bimaterial of piezoelectric materials. *Archive of Applied Mechanics* 68 (6):433-444.

Qin, Q. H., and Y. W. Mai. 1999. A closed crack tip model for interface cracks in thermopiezoelectric materials. *International Journal of Solids and Structures* 36 (16):2463-2479.

Qin, Q. H., Y. W. Mai, and S. W. Yu. 1999. Some problems in plane thermopiezoelectric materials with holes. *International Journal of Solids and Structures* 36 (3):427-439.

Qin, Q. H., C. Qu, and J. Ye. 2005. Thermoelectroelastic solutions for surface bone remodeling under axial and transverse loads. *Biomaterials* 26 (33):6798-6810.

Qin, Q. H., and H. Wang. 2008. *Matlab and C programming for Trefftz finite element methods*: New York: CRC Press.

Qin, Q. H., and H. Wang. 2011. Fundamental solution based FEM for nonlinear thermal radiation problem. *Paper read at 12th International Conference on Boundary Element and Meshless Techniques* (BeTeq 2011), ed. EL Albuquerque, MH Aliabadi, EC Ltd, Eastleigh, UK.

Qin, Q. H., and H. Wang. 2013. Special circular hole elements for thermal analysis in cellular solids with multiple circular holes. *International Journal of Computational Methods* 10 (04):1350008.

Qin, Q. H., and H. Wang. 2015. Special Elements for Composites Containing Hexagonal and Circular Fibers. *International Journal of Computational Methods* 12 (04):1540012.

Qin, Q. H., and J. Q. Ye. 2004. Thermoelectroelastic solutions for internal bone remodeling under axial and transverse loads. *International Journal of Solids and Structures* 41 (9):2447-2460.

Qu, C., Q. H. Qin, and Y. Kang. 2006. A hypothetical mechanism of bone remodeling and modeling under electromagnetic loads. *Biomaterials* 27 (21):4050-4057.

Tao, J., Q. H. Qin, and L. Cao. 2013. A Combination of Laplace Transform and Meshless Method for Analysing Thermal Behaviour of Skin Tissues. *Universal Journal of Mechanical Engineering* 1 (2):32-42.

Wang, H., Y. T. Gao, and Q. H. Qin. 2015. Green's function based finite element formulations for isotropic seepage analysis with free surface. *Latin American Journal of Solids and Structures* 12 (10):1991-2005.

Wang, H., W. Lin, and Q. H. Qin. 2019. Fundamental-solution-based hybrid finite element with singularity control for two-dimensional mixed-mode crack problems. *Engineering Analysis with Boundary Elements* 108:267-278.

Wang, H., and Q. H. Qin. 2010. Fundamental-solution-based finite element model for plane orthotropic elastic bodies. *European Journal of Mechanics-A/Solids* 29 (5):801-809.

Wang, H., and Q. H. Qin. 2011. Fundamental-solution-based hybrid FEM for plane elasticity with special elements. *Computational Mechanics* 48 (5):515-528.

Wang, H., and Q. H. Qin. 2012. Boundary integral based graded element for elastic analysis of 2D functionally graded plates. *European Journal of Mechanics-A/Solids* 33:12-23.

Wang, H., and Q. H. Qin. 2012. A fundamental solution-based finite element model for analyzing multi-layer skin burn injury. *Journal of Mechanics in Medicine and Biology* 12 (05):1250027.

Wang, H., and Q. H. Qin. 2012. A new special element for stress concentration analysis of a plate with elliptical holes. *Acta Mechanica* 223 (6):1323-1340.

Wang, H., and Q. H. Qin. 2012. Numerical implementation of local effects due to two-dimensional discontinuous loads using special elements based on boundary integrals. *Engineering Analysis with Boundary Elements* 36 (12):1733-1745.

Wang, H., and Q. H. Qin. 2015. A new special coating/fiber element for analyzing effect of interface on thermal conductivity of composites. *Applied Mathematics and Computation* 268:311-321.

Wang, H., and Q. H. Qin. 2017. Voronoi Polygonal Hybrid Finite Elements with Boundary Integrals for Plane Isotropic Elastic Problems. *International Journal of Applied Mechanics* 9 (3):1750031.

Wang, H., and Q. H. Qin. 2019. Voronoi Polygonal Hybrid Finite Elements and Their Applications. In *Current Trends in Mathematical Analysis and Its Interdisciplinary Applications*: Springer.

Wang, H., Q. H. Qin, and C. Y. Lee. 2019. n-sided polygonal hybrid finite elements with unified fundamental solution kernels for topology optimization. *Applied Mathematical Modelling* 66:97-117.

Wang, H., Q. H. Qin, and X. P. Liang. 2012. Solving the nonlinear Poisson-type problems with F-Trefftz hybrid finite element model. *Engineering Analysis with Boundary Elements* 36 (1):39-46.

Wang, H., and Q. H. Qin. 2007. Some problems with the method of fundamental solution using radial basis functions. *Acta Mechanica Solida Sinica* 20 (1):21-29.

Wang, H., and Q. H. Qin. 2008. Meshless approach for thermo-mechanical analysis of functionally graded materials. *Engineering Analysis with Boundary Elements* 32 (9):704-712.

Wang, H., and Q. H. Qin. 2009. Hybrid FEM with fundamental solutions as trial functions for heat conduction simulation. *Acta Mechanica Solida Sinica* 22 (5):487-498.

Wang, H., and Q. H. Qin. 2010. FE approach with Green's function as internal trial function for simulating bioheat transfer in the human eye. *Archives of Mechanics* 62 (6):493-510.

Wang, H., and Q. H. Qin. 2010. Fundamental solution-based hybrid finite element analysis for non-linear minimal surface problems. In *Recent Developments in Boundary Element Methods: A Volume to Honour Professor John T. Katsikadelis*, edited by E. J. Sapountzakis. Southampton: WIT Press.

Wang, H., and Q. H. Qin. 2011. A fundamental solution based FE model for thermal analysis of nanocomposites. *Paper read at Boundary elements and other mesh Reduction methods XXXIII,' 33rd International Conference on Boundary Elements and other Mesh Reduction Methods, at UK.*

Wang, H., and Q. H. Qin. 2011. Special fiber elements for thermal analysis of fiber-reinforced composites. *Engineering Computations* 28 (8):1079-1097.

Wang, H., and Q. H. Qin. 2012. Computational bioheat modeling in human eye with local blood perfusion effect. In *Human Eye Imaging and*

*Modeling*, edited by E. Y. K. Ng, J. H. Tan, U. R. Acharya and J. S. Suri. Boca Raton: CRC Press.

Wang, H., and Q. H. Qin. 2013. Fracture analysis in plane piezoelectric media using hybrid finite element model. *Paper read at 13th International Conference of fracture, at Beijing.*

Wang, H., and Q. H. Qin. 2013. Implementation of fundamental-solution based hybrid finite element model for elastic circular inclusions. *Paper read at Proceedings of the Asia-Pacific Congress for Computational Mechanics, at Singapore.*

Wang, H., Q. H. Qin, and Y. Xiao. 2016. Special n-sided Voronoi fiber/matrix elements for clustering thermal effect in natural-hemp-fiber-filled cement composites. *International Journal of Heat and Mass Transfer* 92:228-235.

Wang, H., Q. H. Qin, and W. Yao. 2012. Improving accuracy of opening-mode stress intensity factor in two-dimensional media using fundamental solution based finite element model. *Australian Journal of Mechanical Engineering* 10 (1):41-51.

Wang, J. S., and Q. H. Qin. 2007. Symplectic model for piezoelectric wedges and its application in analysis of electroelastic singularities. *Philosophical Magazine* 87 (2):225-251.

Xu, Y., and D. Chung. 2000. Effect of sand addition on the specific heat and thermal conductivity of cement. *Cement and concrete research* 30 (1):59-61.

Yu, S. W., and Q. H. Qin. 1996. Damage analysis of thermopiezoelectric properties: Part I—crack tip singularities. *Theoretical and Applied Fracture Mechanics* 25 (3):263-277.

Zhang, Z. W., H. Wang, and Q. H. Qin. 2012. Transient bioheat simulation of the laser-tissue interaction in human skin using hybrid finite element formulation. *Molecular & Cellular Biomechanics* 9 (1):31-53.

Zhang, Z. W., H. Wang, and Q. H. Qin. 2014. Analysis of transient bioheat transfer in the human eye using hybrid finite element model. *Paper read at Applied Mechanics and Materials.*

Zhang, Z. W., H. Wang, and Q. H. Qin. 2014. Method of fundamental solutions for nonlinear skin bioheat model. *Journal of Mechanics in Medicine and Biology* 14 (4):1450060.

Zhou, J. C., K. Y. Wang, and P. C. Li. 2019. Hybrid fundamental solution based finite element method for axisymmetric potential problems with arbitrary boundary conditions. *Computers & Structures* 212:72-85.

Zou, M., B. Yu, D. Zhang, and Y. Ma. 2003. Study on optimization of transverse thermal conductivities of unidirectional composites. *J. Heat Transfer* 125 (6):980-987.

# INDEX

## A

advection upstream splitting method, 77
aerodynamic, 76
aero-thermal, 76, 77
algorithm, 18, 61, 62

## B

baseline, viii, 74, 75, 78, 79, 83, 84, 89, 90, 91, 92, 93, 95, 96, 99, 100, 101, 103, 104, 107, 109, 110, 119
boundary value problem, 124, 127, 131, 137
bow shock, 88

## C

cell centroid gradient, 99
cellular materials, 124, 137
chemical, 2, 20, 22, 36
clustering, 142, 151, 152, 153, 163
composites, vii, ix, 123, 124, 125, 142, 155, 157, 158, 161, 162, 163, 164
composition, 2, 20, 22, 25, 36, 69
compressibility, 75, 102, 103, 119
compressibility effects, 75, 83, 84, 102, 103
computational fluid dynamics, 74, 76, 77, 78, 83, 120, 121
computational modeling, 92
computational models, 75, 78
computationally modeling, 78
conduction, vii, viii, ix, 1, 2, 9, 27, 28, 48, 67, 68, 71, 73, 75, 76, 77, 82, 99, 104, 123, 124, 125, 128, 138, 143, 155, 158, 159, 162
conductivity, 2, 3, 4, 7, 14, 21, 25, 36, 58, 60, 62, 63, 68, 69, 70, 71, 82, 83, 87, 126, 139, 141, 142, 144, 149, 150, 153, 157, 158, 159, 161, 163
configuration, 80, 93, 95
conjugate heat transfer, v, vii, viii, 73, 74, 75, 76, 77, 78, 79, 88, 90, 93, 107, 120, 121
conjugate heat transfer conditions, 75
conjugate heat transfer problem, 77, 107
conservation, 2, 25, 81, 82
constituents, 143, 145
construction, 125, 137, 142

contour, 88, 90, 93, 95, 99, 100, 101, 107, 115, 116
contour plots, 88, 90, 93, 95, 99, 100, 101, 107, 115, 116
convergence, 18, 62, 149, 150, 156
cooling, 5, 7, 12, 14, 15, 18, 19, 21, 22, 25, 32, 33, 37, 67, 68, 69
coupled solver, 77, 83
cylinder, v, vii, viii, 2, 5, 28, 31, 44, 67, 73, 74, 75, 76, 77, 78, 79, 80, 81, 82, 83, 84, 85, 87, 88, 89, 90, 93, 94, 96, 98, 99, 101, 104, 105, 106, 109, 110, 115, 116, 119, 120

## D

density based, 77
derivatives, 92, 126, 139
deviation, 17, 18, 22, 23, 24, 25, 26, 47, 62, 107, 153
differential equations, 92
diffusivity, viii, 2, 4, 7, 9, 14, 16, 17, 18, 20, 21, 25, 36, 61, 62, 63, 68
discretization, viii, 74, 75, 77, 83, 84, 92, 93, 95, 96, 110, 119
displacement, 34, 35, 57, 110, 111, 115, 127, 148
distribution, 76, 78, 84, 85, 88, 89, 90, 95, 99, 101, 104, 119, 142, 145, 149, 153
divergence, 132, 147

## E

electromagnetic, 160
elements, ix, 74, 80, 81, 84, 127, 130, 137, 138, 142, 145, 146, 150, 151, 152, 155, 156, 158, 159, 160, 161, 162, 163
energy, 2, 14, 15, 75, 81, 82, 87, 111, 116
energy input, 87
engineering, 124, 137, 156
enhanced wall treatment, 84, 107, 109
equilibrium, 77, 84, 107, 108, 109, 146, 148

## F

fiber, vii, ix, 123, 125, 142, 143, 144, 145, 146, 148, 149, 150, 151, 152, 153, 154, 155, 158, 161, 162, 163
finite element method, vii, ix, 123, 124, 130, 148, 157, 158, 159, 160, 164
finite volume, 78, 83
first order upwinding, 84, 96, 97, 99, 119
flow, viii, 2, 3, 7, 13, 14, 15, 35, 70, 71, 74, 75, 76, 77, 78, 79, 81, 82, 84, 88, 89, 90, 91, 92, 93, 95, 96, 99, 100, 101, 102, 103, 104, 107, 110, 115, 116, 119, 120, 121, 156
flow field, 77, 89, 101, 103, 110, 119
fluid, viii, 68, 73, 75, 76, 77, 78, 79, 80, 81, 84, 90, 93, 110, 111, 116, 119, 120, 121, 156
fluid-solid interface, viii, 73, 75, 76, 77, 84, 90, 119
food, viii, 1, 2, 3, 4, 5, 6, 7, 68, 69, 70
forward facing direct impingement, 88
Fourier series, vii, viii, 1, 2, 5, 9, 28, 39, 40, 68, 71
freedom, 81, 130, 146
fruits, vii, 1, 9, 18, 27, 48, 55, 68, 69, 71
fully turbulent flow, 107
functionally graded material, vii, ix, 123, 124, 125, 139, 141, 155, 158, 162
fundamental solution, v, vii, ix, 123, 124, 127, 128, 134, 135, 136, 137, 139, 140, 141, 142, 145, 146, 148, 155, 157, 158, 159, 160, 161, 162, 163, 164

## G

geometry, 2, 23, 27, 47, 67, 69, 79, 84
gradient limiter, viii, 74, 101, 102, 119
gradient method, 100

graph, 14, 16, 58, 60
green gauss node based, 84, 100, 101

## H

heat conduction, v, vii, viii, ix, 2, 48, 67, 68, 71, 74, 76, 77, 82, 99, 104, 123, 124, 125, 128, 138, 143, 155, 158, 162
heat flux, viii, 73, 76, 77, 84, 85, 86, 87, 89, 90, 93, 94, 96, 98, 99, 101, 104, 105, 106, 107, 110, 111, 112, 114, 115, 116, 118, 119, 120, 126, 139, 144, 146, 147
heat flux distribution, 84, 85, 90, 119
heat of respiration, 2, 9, 28, 37, 39, 46, 48, 63, 71
heat transfer, vii, viii, 1, 2, 3, 4, 7, 8, 9, 15, 17, 18, 22, 25, 27, 28, 35, 48, 58, 59, 62, 67, 68, 69, 73, 74, 75, 76, 77, 78, 79, 87, 88, 90, 93, 106, 107, 120, 121, 144, 149, 157, 158
heating rate, 71
hemp, vii, ix, 123, 125, 142, 143, 145, 148, 149, 150, 151, 152, 154, 155, 157, 159, 163
hemp fiber, 142, 143, 145, 148, 149, 150, 151, 152, 154, 157, 159
heterogeneity, 3
high speed compressible, v, vii, viii, 73, 74, 75, 77, 90, 119, 120
high speed compressible flow, v, vii, viii, 73, 74, 75, 77, 90, 119, 120
high speed flow, 78, 79, 88, 110, 119, 121
high speed gas flow, 78
history, 16, 39, 40, 41, 42, 58
hybrid, 127, 128, 140, 145, 148, 155, 157, 158, 161, 162, 163
hypersonic conditions, 77
hypersonic flow, 76, 88

## I

interface, 15, 16, 18, 25, 48, 49, 64, 75, 76, 77, 84, 85, 87, 90, 93, 95, 104, 111, 115, 119, 120, 143, 144, 159, 160, 161
isotropic media, 144

## L

least square cell based, 99
least square cell based gradient, 99
linear function, 48
linear model, 28, 35
logarithmic coordinates, 16, 58, 60

## M

mach number, 79, 88, 115, 116
mass, vii, viii, 2, 11, 12, 15, 21, 25, 26, 32, 33, 36, 38, 41, 52, 68, 81
materials, vii, viii, ix, 70, 73, 123, 124, 125, 137, 140, 141, 155, 156, 157, 158, 159, 160, 162
matrix, 126, 127, 133, 139, 141, 142, 143, 144, 145, 146, 148, 149, 150, 151, 154, 155, 163
microstructure, 142, 156
modeling methods, viii, 74, 75, 77, 78, 79, 83, 90, 93, 110, 119
models, viii, 9, 14, 27, 69, 74, 75, 76, 78, 83, 104, 106, 107, 110, 115, 120, 141, 155
modifications, 8, 84, 107, 119
momentum, 81, 115, 116
moved, 79, 83
moving, viii, 25, 73, 76, 78, 79, 81, 83, 88, 101, 104, 110, 111, 115, 116
moving cylinder, viii, 74, 78, 81, 83, 110
multidimensional, 101, 121

## N

near sonic, ix, 74, 79, 110
nodes, 129, 130, 134, 148, 149, 150, 151, 152
non-equilibrium wall treatment, 107, 109
numerical analysis, 135
numerical tool, 156

## O

optimization, 125, 156, 162, 164
overset, ix, 74, 80, 81, 111, 113, 115, 116, 117, 120
overset mesh, ix, 74, 81, 115, 120
overshoots, 101

## P

physical properties, 3, 7
piezoelectricity, 158
porous materials, 69
position dependent, 91
potato, 28, 35, 36, 37, 41
pressure, ix, 74, 75, 76, 77, 78, 79, 83, 84, 85, 88, 89, 90, 91, 92, 93, 94, 95, 97, 98, 99, 100, 101, 102, 103, 104, 105, 106, 107, 108, 109, 110, 112, 114, 115, 116, 118, 119, 120, 121
pressure gradient, 84, 107, 121
pulp, vii, 1, 9, 15, 16, 18, 19, 25, 26, 48, 49, 50, 57, 58, 60, 63, 64

## R

radial and angular variation, 76
radius, 8, 9, 19, 25, 35, 48, 63, 136, 148, 149
recommendations, iv
reconstruction, 121
refinement, 84, 86, 115, 120, 156
regression, 16, 17, 19, 21, 26, 27, 28, 29, 58, 59, 60, 64, 66
regression line, 19, 21
regression model, 27
respiration, 2, 8, 9, 27, 28, 35, 37, 39, 41, 46, 47, 48, 53, 58, 63, 71
roots, 6, 22, 65

## S

second order upwinding, 83, 84, 96, 99, 101, 119
seed, vii, 1, 9, 10, 48
sensitivity, vii, ix, 74, 75, 78, 84, 87, 96, 116, 119, 120
shape, 3, 8, 19, 23, 48, 77, 104, 110, 131, 146, 150
shock, 76, 77, 84, 88, 90, 110, 116, 121
shock boundary, 84
shock interface, 84
shock structure, 88
simulation, viii, 70, 74, 90, 101, 150, 162, 163
simulations, 75, 83, 110, 121
six-degree-of-freedom, 81
skin, 161, 164
sliding, viii, 74, 80, 111, 113, 115, 116, 117, 120
sliding mesh, 80, 111, 115, 116, 120
solid cylinder, 78, 79, 82, 87, 88, 99, 111
solution, vii, viii, ix, 1, 3, 5, 9, 11, 27, 28, 30, 49, 56, 68, 71, 96, 101, 103, 115, 119, 123, 124, 127, 128, 129, 134, 135, 136, 137, 140, 141, 142, 145, 155, 157, 158, 160, 161, 162, 163, 164
specific heat, 2, 4, 7, 21, 25, 36, 82, 83, 87, 163
stagnation, 75, 76, 77, 84, 86, 87, 88, 91, 95, 99, 104, 107, 109, 116
stagnation point, 76, 104, 110

stagnation pressure, 110, 116
stagnation zone, 75, 88, 91, 96, 107, 109
standard deviation, 23
standard error, 22, 66
standard κ–ω turbulence, 104
state, 27, 58, 59, 82, 143, 157, 158
steel, ix, 74, 76, 77, 82, 83, 87
stone fruit, v, vii, 1, 2, 9, 47, 48, 49, 50, 55, 68
storage, 18, 27, 62, 68, 71, 87
stress, 82, 161, 163
stress intensity factor, 163
stretching, 81
structural gene, 69
structure, 88, 156
subdomains, 127
subsonic, ix, 74, 79, 110
substitution, 26, 133
Sun, 121
supersonic, ix, 74, 79, 110
susceptibility, 69
symmetry, 3, 4, 133, 137

## T

techniques, 83, 92, 99, 110, 116
temperature, vii, viii, 1, 2, 3, 4, 6, 7, 10, 14, 15, 16, 18, 22, 25, 26, 27, 28, 31, 32, 33, 34, 35, 37, 38, 39, 40, 41, 42, 43, 45, 47, 48, 50, 51, 52, 53, 56, 57, 58, 59, 60, 62, 64, 66, 67, 68, 69, 71, 73, 75, 76, 78, 79, 82, 87, 88, 89, 90, 91, 92, 93, 94, 95, 97, 98, 99, 100, 101, 102, 103, 104, 105, 106, 107, 108, 109, 110, 111, 112, 113, 114, 115, 116, 117, 118, 119, 120, 126, 136, 138, 143, 144, 145, 150
temperature dependent properties, 76
temperature field, 75, 78, 79, 88, 90, 99, 101, 104, 111, 115, 119, 126, 143, 145, 150
thermal analysis, 160, 162
thermal conditions, viii, 74, 75, 76, 78, 87, 92, 93, 100, 104, 106, 107, 110, 119, 120
thermal energy, 75, 116
thermal properties, 69, 70, 77, 87, 142
thermal resistance, 5
time discretization, viii, 74, 83, 84, 92, 93, 96, 119
timestep, 75, 83, 90, 93, 95, 96, 99, 119
topology, 125, 156, 162
total energy, 13
transient heat transfer, 2, 67
treatment, 25, 75, 84, 107, 109, 121
trial, 17, 44, 61, 62, 127, 137, 155, 156, 162
turbulence, viii, 74, 75, 77, 78, 83, 84, 102, 103, 104, 105, 106, 107, 108, 109, 119, 121

## U

upwinding, 84, 96, 99

## V

variables, x, 4, 10, 28, 123
variations, 103
vector, 96, 126, 133, 134, 139, 146
vegetables, 27, 69, 71
velocity, 75, 79, 83, 90, 101, 107, 110, 111, 115, 116, 119, 120
viscous sublayer, 107

## W

wake region, 76, 88, 91, 95, 99, 104, 109, 115
water, 12, 18, 25, 62, 64, 71

wave propagation, 124, 159

## Y

$y^+$, 108